AF556003

Community Ecology of Tropical Birds

Plate 1

Black Ibis - *Pseudibis papillosa* (Temminck)

Black-winged Stilt - *Himantopus himantopus* (Linnaeus)

Plate 2

Brahminy Kite - *Haliastur indus*

Bronze-winged Jacana - *Metopidius indicus* (Latham)

Plate 3

Common Hoopoe - *Upupa epops* (Linnaeus)

Common Redshank - *Tringa totanus*

Plate 4

Common Sandpiper - *Actitis hypoleucos* (Linnaeus)

Curlew Sandpiper - *Calidris ferruginea* (Pontoppidan)

Plate 5

Emerald Dove - *Chalcophaps indica* (Linnaeus)

Eurasian Curlew - *Numenius arquata* (Linnaeus)

Plate 6

Eurasian Golden Oriole - *Oriolus oriolus* (Linnaeus)

Grey-headed Flycatcher - *Culicicapa ceylonensis* (Swainson)

Plate 7

Little Egret - *Egretta garzetta* (Linnaeus)

Oriental White Ibis - *Threskiornis melanocephalus* (Latham)

Plate 8

Painted Stork - *Mycteria leucocephala* (Pennant) 1

Painted Stork - *Mycteria leucocephala* (Pennant) 2

Plate 9

Painted Stork - *Mycteria leucocephala* (Pennant) 3

Purple Moorhen - *Porphyrio porphyrio* (Linnaeus)

Plate 10

Red-wattled Lapwing - *Vanellus indicus* (Boddaert)

Spot-billed Duck - *Anas poecilorhyncha*

Plate 11

Whiskered Tern - *Chlidonias hybridus* (Pallas) 1

Yellow-legged Gull - *Larus cachinnans* (Pallas) 2

Plate 12

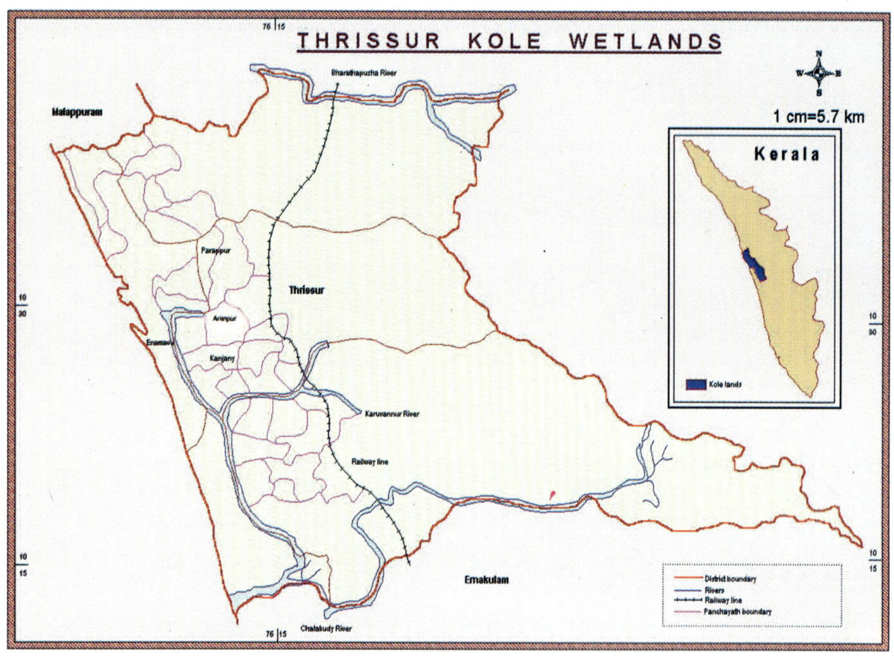

Kole Map Recent

Kole Wetland 1

Plate 13

Kole Wetlands 2

Kole Wetlands 3

Plate 14

Kole Wetlands 4

Kole Wetlands 5

Plate 15

Kole Wetlands 6

Silent Valley 1

Plate 16

Silent Valley 2

Silent Valley 3

Plate 17

Silent Valley 4

Silent Valley 5

Plate 18

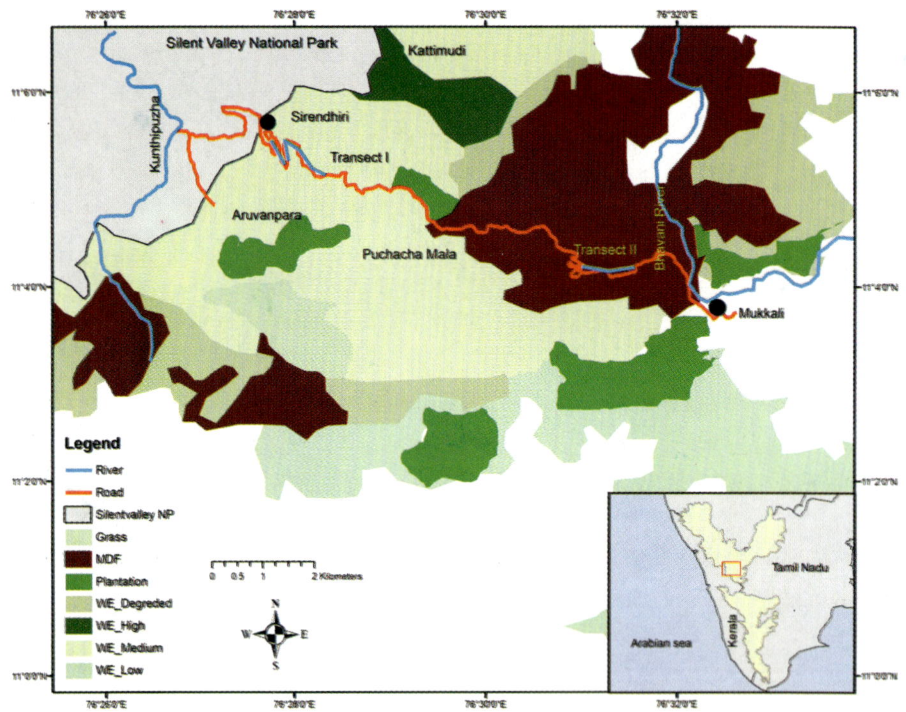

Silent Valley Map

❐❐❐

COMMUNITY ECOLOGY OF
Tropical Birds

Authors

Dr. E.A. JAYSON
Division of Wildlife Biology
Kerala Forest Research Institute
Peechi - 680 653, Kerala, India

Dr. C. SIVAPERUMAN
Zoological Survey of India
Andaman & Nicobar Regional Centre
Port Blair - 744 102
Andaman & Nicobar Islands, India

2010

New India Publishing Agency
Pitam Pura, New Delhi-110 088

Published by
Sumit Pal Jain *for*
New India Publishing Agency
101, Vikas Surya Plaza, CU Block, L.S.C. Mkt.,
Pitam Pura, New Delhi- 110 088, (India)
Phone: 011-27341717, Fax: 011-27341616
Mobile : 09717133558
E-mail: newindiapublishingagency@gmail.com
Web: www.bookfactoryindia.com

ISBN : 978-93-80235-16-5

Typeset at: **Harminder** *for* **Laxmi Art Creations** # 98 11 48 23 28
Printed at: Jai Bharat Printing Press, Delhi

Foreword

Tropical ecosystems are one of the most diverse biological habitats on earth. Seventy six per cent of all centers of avian endemism occur in tropical regions and the same is true for many plant and animal communities. Birds are an important element of biological diversity and their ecological, cultural, recreational and economic benefits are recognized universally. Healthy bird populations are an integral component of pristine ecosystems. They act as vital links in many food webs and often serve as highly visible biological indicators of ecosystem health. A wide range of bird populations are declining all over the world due to habitat loss and fragmentation, predation, pesticide use, invasive exotic species and other factors.

Tropical avian communities have been an attractive study model for ecologists around the globe. Many of the initial efforts focused on the patterns and processes at a local scale. Some of these attempts fell short because of a limited perspective and due to the tendency to focus on single variables. Tropical bird communities are an interesting mix of stable and dynamic systems with relatively sedentary populations as well as some migratory populations following temporal and spatial changes in food resources. Thus, recent studies have emphasized the necessity of multi-scale, multi-factor approaches and the inclusion of both short-term and long-term temporal variables. The difficulties of assessing numerous variables and various scales seem to have deterred ecologists in recent years from embarking on studies of avian communities. However, there is still a pressing need to study these communities at an all-inclusive, composite level, both for understanding basic ecology and for enlightened land management and conservation.

This publication on the community ecology of tropical birds is therefore, a significant contribution to the avian community. The various chapters on different aspects of tropical forest and wetland bird community are a notable advancement of our current knowledge. The authors of this book have meticulously presented the information in a format that can serve as a ready reference point for future research and I sincerely believe that the scientific fraternity will welcome this valuable resource. The publication of this priceless volume is the outcome of the concerted effort of the authors, Dr. E.A. Jayson and Dr. C. Sivaperuman. Their tireless work and thorough proficiency is commendable.

Place: Said Nagli
J.P. Nagar (UP)
Date: 01st October, 2009

Dr. Qaiser Husain Baqri
Ph.D, D.SC., *FZSI, FNASC, FNAAS, FNSI*
Additional Director (Retd)
Zoological Survey of India

Preface

Forest Bird Community

The community ecology of birds was carried out in Silent Valley National Park and Mukkali areas in the tropical forest of Southern Western Ghats from May 1988 to April 1993. The study area is situated in Palakkad District, Kerala State (11° 3' and 11° 13' N latitude and 76° 25' and 76° 35' E longitude). The major forest types in the area are west coast tropical evergreen forests, subtropical broad leaved hills forests, moist deciduous forest and grass lands. The study on birds was based on direct observation and density of birds was estimated using variable-width line transects in each month. To correlate the diversity of birds to the vegetation in the area, vegetation parameters like percentage composition of trees, vegetation profile diagram and foliage height profile of Silent Valley and Mukkali were prepared. In addition to this, girth class distribution of trees, maturity index of vegetation and diversity of trees were also measured using standard methods.

Total of 137 taxa of birds were recorded from the two areas. A variation in species composition was observed during the different months and a significant increase in the number of species and total number of birds was observed during the summer months. The resident birds showed a stable population while the second group of local migrants registered an increase in abundance during dry months. Species richness indices were high at Mukkali and low at Silent Valley. Species-abundance models at these two places followed truncated lognormal distribution. Proportional abundance indices like Shannon Index, Simpson Index, Hill's numbers showed identical values at both the areas. Shannon Index of the two areas showed significant difference. Similarly indices like Jaccard Index, Sorenson Index and Sorenson quantitative Index showed that the two areas were similar in bird communities only at 40 per cent level.

Rainfall is found to have significant negative correlation on the species richness, than on the total number of birds or density at Silent Valley. Similarly, significant difference is obtained between summer and monsoon season in total number and density of birds at Silent Valley. But no such difference is obtained at Mukkali. A significant positive correlation was obtained between the total number of birds in each month and the abundance of flowers. However, no significant correlation was obtained between total number of birds and other food items, such as insects and fruits.

Bird species used all the seven foliage height categories identified at the Silent Valley and Mukkali. High species richness of birds was observed at lower and middle levels at both the places. Four diversity indices were worked out for seven foliage height categories at the two sites. The diversity was in a decreasing order from the lowest to the highest stratum. At Mukkali, highest diversity was found in the third stratum (6-10). A significant positive correlation was obtained between foliage abundance and total number of birds and also with species richness in each height stratum, at Silent Valley and Mukkali. In all the height strata, omnivorous and insectivorous birds were of equal abundance followed by others.

Foraging ecology observations revealed that birds were using basically 5 foraging methods and gleaning was a principal method used by 13 species followed by probing. No complete General Overlap was found in any of the foraging parameters studied. Insectivorous birds were maximum in both of these areas followed by omnivorous and frugivorous birds. There were more frugivorous birds at Silent Valley than at Mukkali. Microhabitat utilization of birds showed that branches of trees and foliage were highly utilized followed by others. The general overlap in microhabitat use was low in both the study areas. The study analyses, bird community parameters and show its linkage to the habitat, vegetation and climatic parameters. Comparison of the two tropical forest bird communities with different supporting vegetation show the variations and similarities in the parameters studied. The evaluation of the study areas shows the value of this reserve forest, which is comparable to the world standards and recommends its preservation to function as a buffer zone for the Silent Valley National Park.

Wetland Bird Community

This study has been conducted in the Kole lands of Thrissur, Kerala which is part of Vemband-Kole Ramsar site during November 1998 to October 2001 (10° 20' and 10° 40' N latitudes and between 75° 58' and 76° 11' E longitudes), with an extent of 13,632 ha spread over Thrissur and Malappuram Districts, Kerala State. The Kole wetlands are low lying tracts located 0.5 to 1 m below MSL and it remains submerged for about six months in a year. Climate of the area is moderate and three different seasons are found in the study area. The temperature varies from 28° C to 31.5° C and the average annual rainfall is 3,200 mm and the maximum rainfall is received during June and July.

Community parameters of birds were studied based on direct observation and four intensive study areas were selected for detailed observations. Density of birds was estimated using total count method. The food and feeding patterns of five species were studied. In order to compare the availability of invertebrate fauna with the bird community, their abundance in different microhabitats was estimated during three migratory seasons. For this, mud samples were collected from different microhabitats by Naturalist's dredge method and invertebrates were estimated. Macro fauna was collected from shallow water, mud flats and paddy fields using 1 m X 1 m quadrat method. Microhabitat utilisation pattern of 11 selected bird species was studied. Conservation awareness of local people was assessed through structured questionnaire survey by directly interviewing the respondents in the study area. Species richness, abundance, diversity indices, density, seasonal fluctuations of bird community, food and feeding of selected bird species, habitat utilisation and conservation problems of wetland birds were recorded and analysed using statistical packages.

A total of 182 taxa of birds, belong to 50 Families under 16 Orders were recorded. Of the 182 species, 100 were resident, 81 migrants and one straggler. Among the migrants, 49 species were trans-continental migrants and 32 local.

Out of the 182 species, 48 were new records for the area. One vulnerable and five near threatened species were recorded, namely Spot-billed Pelican (*Pelecanus philippensis*), Darter (*Anhinga melanogaster*), Painted Stork (*Mycteria leucocephala*), Oriental White Ibis (*Threskiornis melanocephalus*), Ferruginous Pochard (*Aythya nyroca*), and Pallid Harrier (*Circus macrourus*).

Species richness of birds varied in different months and the highest recorded number of species was 97 during December 1999 and the lowest was 15 during June 1999. The species richness increased during the migratory season and decreased during the southwest monsoon. Total number of birds varied from 35 to 8,033 individuals in a month. The highest number of birds (8,033) was recorded during November 2000 and the lowest (35) during June 1999. The highest diversity Index (H') was recorded in December (3.01) and lowest in October (2.11). The highest density of birds was recorded in December (29,158 birds/ha) followed by November (24,373 birds/ha).

Among the four intensive study areas, species richness was highest at Kanjany (121) followed by Parappur (117), Enamavu (94) and Chettupuzha (71). Monthly analysis showed that the highest number of species was recorded at Parappur during December (79) and the lowest at Chettupuzha during June (10). The maximum number of birds was recorded from Kanjany (73,604) during December and the minimum from Chettupuzha (140) during July. The highest diversity Index (H') was observed in June (2.90) at Parappur and the lowest in February (1.53) at Parappur. The highest density was in December (53,994 birds/ha) at Kanjany and the lowest was in July (103 birds/ha) at Chettupuzha.

Out of the 82 wetland bird species observed, Whiskered Tern (23 per cent) was highest in dominance followed by Little Egret (13 per cent) and Little Cormorant (11 per cent). Species richness and abundance were highest in the dry season of 2000 (120) and lowest in the wet-I season of 1999 (46). Diversity Index (H') was highest in the dry season of 2000 (3.19), followed by the dry season of 2001 (2.89). Species richness was highest in the third year (2000-2001) migratory season (127) followed by second year (1999-2000) migratory season (119). The highest abundance of birds was recorded in the third year migratory season (1,88,006) and lowest in the first year migratory season (19,350). A significant negative correlation was found between rainfall, water depth and bird population parameters. The highest number of birds was recorded during the replanting season and the sowing period of paddy. A significant negative correlation was also found between the paddy height and the abundance of birds.

Shallow water was highly preferred among the 12 microhabitats. Species diversity Index (H') was highest on the trees (3.44) and lowest on the bunds (1.38). Species density was highest in the shallow waters (67,788 birds/ha) followed by paddy fields (56,774 birds/ha) and mud flats (56,308 birds/ha). Highest niche breadth was recorded for Red-wattled Lapwing and Indian Pond-Heron. Mean species richness was significantly higher on the electric lines during dry season, wet-I season and in the mud flats during the wet-II season. Diversity Index (H') was significantly higher on trees in the dry and wet-II season; during the wet-I season, this was higher on the electric lines. Mean density of birds was higher in the paddy fields in the dry and wet-II season and significantly higher in the floating vegetation during wet-I season.

The Kole wetlands showed high species richness and abundance of birds, and are comparable to other wetlands and protected areas in Kerala. The Kole wetlands are an ideal habitat for migratory and resident birds, especially for the winter visitors. In the present study, 49 species of trans-continental migrants were recorded from the Kole wetlands, which showed the importance of the area as a wintering ground for migratory species. This wetlands supported waders similar to the known sites such as Chilika Lake, Pulicate Lake, Great Vedaranyam Swamp, Point Calimere and Gulf of Mannar.

Among the four intensive study sites, highest number of birds was recorded from Kanjany, which can be attributed to the geographic position of the study site, which is in the middle of the Kole wetlands and for the high availability of mud flats. The increase in the wetland bird species from September to March in all the years reflected the availability of ideal microhabitats and higher production of benthic and macro fauna. The rainfall and the water depth were the major climatic factors influencing the abundance of birds at Kole wetlands. The area served many avian species for a wide variety of purposes such as nesting, roosting and wintering ground.

Food and feeding behaviour studies of selected bird species showed that sufficient prey is available in the Kole wetlands and the selection of food depended on the variation in the feeding technique, body size and bill morphology. The study indicated a strong positive correlation between the wader abundance and the benthic fauna. The appearance of mud flats attracted large number of waders during the migratory season.

As the Kole wetlands come under the 'Central Asian - Indian flyway' and one of the Ramsar Sites in India, protection of migratory bird species is of the highest priority.

Thrissur and Port Blair

E.A. Jayson
C. Sivaperuman

Acknowledgement

We express our heartfelt gratitude to all those who helped in different ways to complete this work. Our sincere thanks go to Dr. K.V. Sankaran, The Director and former Director's, Kerala Forest Research Institute, Peechi, Dr. C. Renuka, Programme Coordinator, Division of Forest Ecology and Biodiversity Conservation. Our Sincere thanks to the Department of Forests and Wildlife, Kerala for financial assistance to this study. The second author acknowledges Dr. Ramakrishna, the Director, Zoological Survey of India, Kolkata and Dr. C. Raghunathan, Officer-in-Charge, Zoological Survey of India, Andaman & Nicobar Regional Centre, Haddo, Port Blair for providing necessary facilities. Prof. D.N. Mathew, Rtd. Professor, University of Calicut is acknowledged for the guidance to the first author. Statistical advice was given by Dr. K.A. Mercey, Professor, Dept. of Statistics, College of Veterinary and Animal Sciences, Kerala Agricultural University, Thrissur and Dr. M. Sivarams, Scientist, KFRI, Peechi. Figures were drawn by Shri Subhash Kuriakose, Artist Photographer, KFRI. Many people assisted the work in the field and all are acknowledged for the same.

Contents

PART - I

FOREST BIRD COMMUNITIES

Chapter 1

Introduction

Birds possess great intrinsic interest, they are certainly among the most attractive population of all wildlife groups. Moreover, it is widely recognised that birds can act as valuable indicators of the quantity and quality for wildlife. The studies on avian communities attracted great attention over the past several years all over the world. Much of the early efforts emphasized on the patterns and processes in the avian communities and the listing of species. Previous workers mainly concentrated on the narrow perspectives and focused on single variables, but the recent studies have emphasised the necessity of multi-scale, multi-factor approaches and the inclusion of both short term and long-term temporal variables. The difficulties of assessing numerous variables and several scales, both temporal and spatial, seem to have discouraged the contemporary ecologists from embarking on studies on avian communities.

The Western Ghats is well known for the wealthy and unique assemblage of flora and fauna and is one among the 25-biodiversity hot spots identified on the earth. Arising gradually from the narrow

Konkan and Malabar coasts, these mountains run 1,600 km north south between the River Tapti in Gujarat and Kanyakumari in Tamil Nadu covering an area approximately equal to 1,60,000 km^2. Kerala State, located between 8° 4' and 12° 48' N and 74° 52' and 77° 37' E is known for rich biological resources on account of availability of a variety of ecological niches and habitats ranging from lush tropical forest, valleys, plains and coastal areas.

Tropical Forest Bird Studies in Kerala

The Kerala State is rich in bird fauna and a total of about 493 species have been recorded from the State (Jayson and Sivaperuman, 2006). The scientific studies on the birds of Kerala commenced with A.O. Hume (1876 and 1878) reporting the first and second list of birds of Travancore in the southern Kerala and followed by Bourdillon (1880), Ferguson and Bourdillon (1903, 1904a, 1904b and 1904c). In northern Kerala Prime Rose (1904) reported on birds in the Nilgiris and Wynaad; and Baker (1911) on birds in Kannur.

The golden era of ornithology in Kerala started with the ornithological survey of Travancore and Cochin by Sálim Ali. He published the results of the survey in 8 volumes (Ali and Whistler, 1935-1937). The "Birds of Kerala" published in 1969 by Sálim Ali is still the authentic record on birds in the State. During the fifties, Neelakantan initiated studies on the birds of Kerala and published the book in Malayalam entitled "Keralathilae Pakshikal" (Neelakantan, 1958) and he has published several papers on various aspects of birds (Neelakantan, 1968 and 1990). Jackson (1954a and 1954b) published many new records of the birds in Kerala.

During the 1980's researchers from the Calicut University, studied the exhaustive ecology of the individual species of birds and the community ecology under the guidance of Prof. D.N. Mathew. Zacharias (1979) worked on Babblers, Ramakrishnan (1983) on the birds of Malabar forest and Jayson and Mathew (2000a, 2000b, 2002, 2003) on the community ecology of birds in the Silent Valley. Among the other studies Satheesan (1990) worked on Pariah Kite, Yahya (1980) on Barbets, Vijayan (1984) on Drongos and Santharam (1995) on Woodpeckers. Zacharias and Gaston (1999) made a detailed survey in the Western Ghats from 1973 to 1997 and reported 309 species of birds including 67 winter visitors. Robin and Davidar (2002) studied the vertical stratification of mixed feeding flocks in the moist deciduous forest and teak plantations at Parambikulam Wildlife Sanctuary.

Wetlands of Kerala

Wetlands in Kerala are distributed all along the coast and in the inlands. Prominent coastal wetlands in Kerala are Vellayani Kayal, Aakkulam-Veli backwater stretch, Kayamkulam Pozhi, Kumarakam, Mangalavanam, Kole wetlands, Purathur estuary, Manaloor Kayal, Chervarpur Kayal, Kadalundy estuary, Azhinijilam, Dharmadom estuary, Kattampalli, Ezhimala, Chempallikundu and Mangrove areas (Kurup, 1996). The important fresh water wetlands are Sasthamkotta Lake, Pookot Lake, and Muriyad. Wetlands in Kerala are under extreme pressure due to the high population density of the State. According to Gopalan (1991), as much as two-third area of Vembanad Lake has been either reclaimed as land or converted into fields for agricultural and fishery activities. Wetlands in Kerala are mainly used for agriculture, aquaculture, reclamation for housing and industrial purposes, disposing the waste materials, discharging the industrial effluents and municipal waste water, wood seasoning, feeding waters for ducks, dumping dredged soil, coir retting and for fishing (Balachandran *et al.*, 2002).

Water birds are important components of most of the wetland ecosystems, as these occupy several trophic levels in the food web of wetland nutrient cycles. Water birds are broadly defined as 'birds ecologically dependent on wetlands' and include recognised groups, popularly known as wild fowls, waterfowls, shorebirds or waders. In addition to these groups, other groups dependent on wetlands are passerines. Several wetlands in the coastal floodplains are important for the migratory waders and ducks. As the shorebirds use varied habitats like estuaries, riverbanks, paddy fields, and shallow water for foraging and sites for roosting are readily available. In the Asia-Pacific region, 243 species, by virtue of their nature, undertake annual migrations between the breeding areas and non-breeding grounds, along various flyways. Wetlands in Kerala come under Central Asian-Indian flyway (Anon., 1996). During the annual migration, water birds halt at sites for very short periods to rest and feed and these 'stepping stones' are essential for their survival. Many species of wetland birds also play a role in control of agricultural pests, while some species are themselves considered pests of paddy.

Studies on Tropical Wetland Birds in Kerala

The ornithology of Kerala wetlands received attention after Neelakantan's extensive explorations (Neelakantan, 1969 and 1970; Neelakantan *et al.*, 1981; Neelakantan and Sureshkumar, 1981).

Uthaman and Namasivayam (1991) explored the birds of Kadalundy. Ravindran (1993, 1999 and 2001) had reported the occurrence of Glossy Ibis (*Plegadis falcinellus*), Whitenecked Stork (*Ciconia episocopus*) and White-winged Black Tern (*Chlidonias leucopterus*). Ravindran (1995) and Jayson and Sivaperuman (1999) have highlighted the importance of conserving the wetlands birds in the Kole wetlands.

Some of the important recording of wetland bird species from various parts of Kerala is given below. Ambedkar (1985) reported the Sandwich Tern (*Sterna sandvicensis*) from Ernakulam District. Least Frigate Bird (*Fregata ariel*) was recorded by Faizi (1985) from Quilon. Jairaj and Kumar (1990) reported the Spoonbill (*Platalea leucorodia*) from Thrissur. Black-tailed Godwit (*Limosa limosa*) and Large Indian Pratincole (*Glareola pratincola*) were recorded by Kumar (1990) from Baliapattam River, Kannur. Namasivayam and Venugopalan (1990) reported Avocet (*Recurvirosta avocetta*) from Kadalundi Estuary and Spotbill Duck (*Anas poecilorhyncha*) was recorded by Uthaman (1990) from Mangalam Reservoir, Palakkad and Water Cock (*Gallicrek cinerea*) by Ray (1992) from Kuttanad.

Sashikumar (1992) reported Indian Shag (*Phalacrocorax fuscicolis*) from Kattampally. Sathasivam (1992) recorded Painted Stork (*Mycteria leococephala*) from Periyar Tiger Reserve. Ravindran (1994 and 1998) reported Gadwall (*Anas strepera*) from Kadalundy Estuary, Black-bellied Tern (*Sterna acuticauda*) from Palakkad and Comb Duck (*Sarkidiornis melanotus*) from Malappuram District. Kumar and Kumar (1996) reported Christmas Frigate Bird (*Fregata andreusi*) from Malappuram. Glossy Ibis was identified by Praveen and Kumar (1996) from Palakkad and Lesser Frigate Bird (*Fregata minor*) was recorded by Saraswathy (1999).

Kurup (1991) studied the ecology of wetland birds in the Malabar Coast and Lakshadeep and Uthaman and Namasivayam (1991) described the bird life of Kadalundi Estuary. Inventory of birds from Vembanad Lake were also reported during the period (Anon., 2001 and 2003). Jayson and Easa (2000) carried out a detailed study on the birds of Mangalavanam mangroves and Chellappan & Thomas (2001) described the avifauna of rice fields of Pattambi, Palakkad District. Jayson (2000), Sivaperman and Jayson (2002a), Jayson and Sivaperman (2003) reported Black Stock (*Ciconia nigra*), Northern Shoreller (*Anas clypeata*) and Lesser Frigate Bird (*Fregate ariel*) from the kole wetlands. The Structure, Species composition, diversity, popolation dynamics and habitat utilization of birds of kole. Wetlands were reported by sivaperman & Jayson (2001a, 2001b, 2002b & 2003).

STUDY AREA

Silent Valley National Park

Location and Topography

The study area is located in Palakkad District, Kerala State, India between latitude 11° 3' and 11° 13' N longitude and 76° 25' and 76° 35' E is 45 km, North of Mannarghat, the nearest town, in the Western Ghats of South India. The area comprises the Silent Valley and Attappady reserve forest (Fig. 1.1). The former reserve is on a Plateau and its northern side is contiguous with Nilambur and Nilgiris. The southern portion is bordered by the Palakkad vested forests and the western limits are also covered by the vested forests of the Nilambur Division.

Two intensive study sites were selected in the study area; one was near the abandoned dam site at Silent Valley and the other at Mukkali. The study site at Silent Valley lies adjacent to the core area of Nilgiri Biosphere Reserve. The study site at Mukkali is in the Attappady Reserve Forest and lies near Mukkali Village.

The first study site is partially degraded to some extent and most of the disturbance happened in the late seventies and early eighties, in the course of felling trees and pre-construction work of the abandoned dam. But after the declaration of the Silent Valley National Park in 1984, this tract received maximum protection from poaching and fire.

Evergreen Forest of Silent Valley National Park

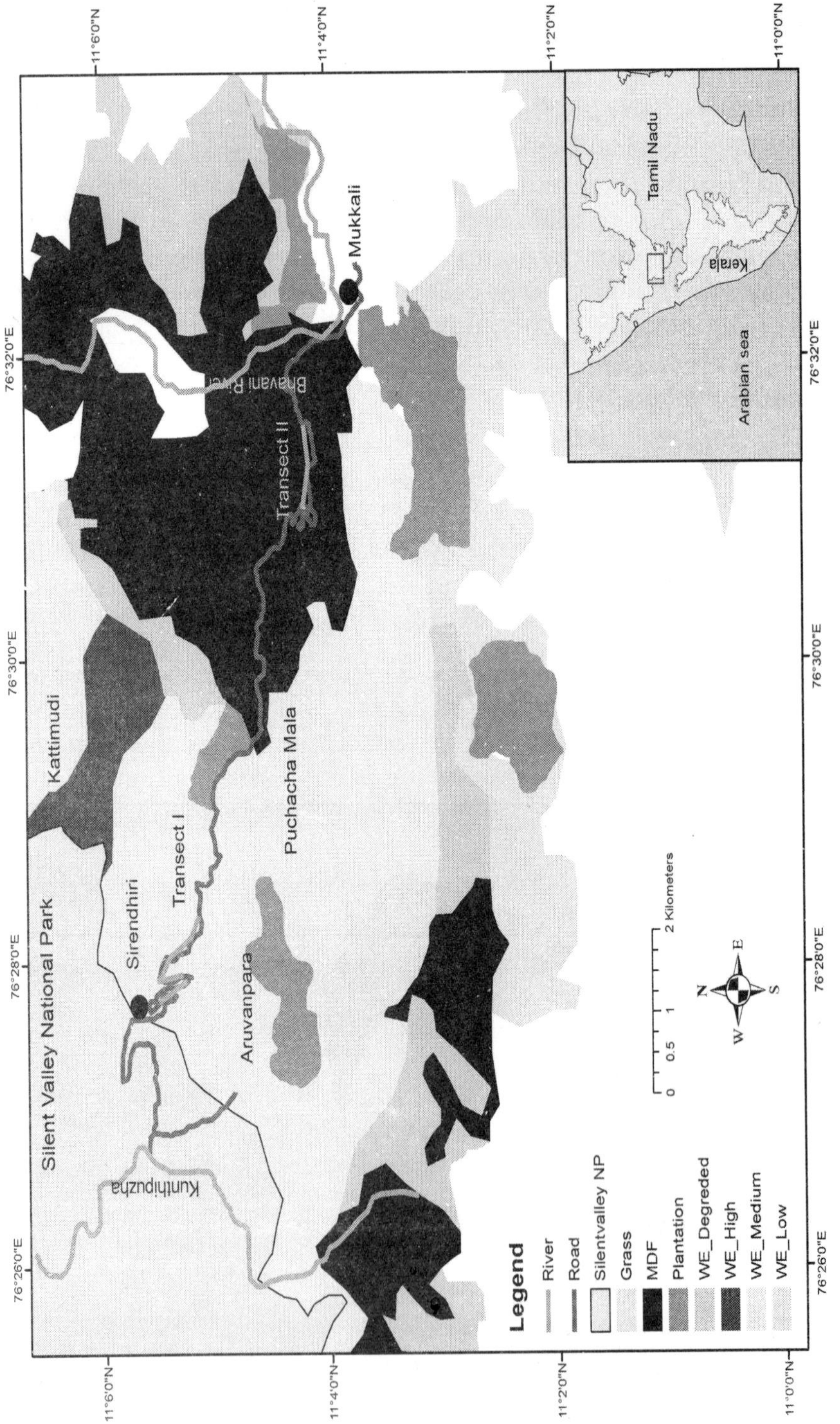

Fig. 1.1 : Silent Valley National Park of Mukkali Reserve Forest.

Silent Valley

The main drainage of this area is through the Kunthipuzha River, which is the tributary of the Bharatapuzha. The elevation varies from 500 m to 1500 m and topography is undulating. Poochipara (1,376 m) and Anginda (2,383 m) are some of the peaks. As per the world classification of Udvardy (1975) Silent Valley falls under the Malabar Rain Forest Realm. According to the Biogeographic classification of Rodger and Panwar (1989) the area falls under the Biographic zone 5 Western Ghats and Biogeographic Province 5B Western Ghats Mountains and Biogeographic Sub-division Nilgiri.

Mukkali

Comes under the buffer zone of the Nilgiri Biosphere Reserve in Attappady Reserve Forest. The Attappady reserve lies on the eastern side of Silent Valley. The main drainage is through the Bhavani River, which takes its origin from the Nilgiris. Northern side is bordered by Nilgiris, East and South by private lands. Malleswara Malai, about 1681 m, is one of the highest peaks. The eastern side of the Attappady reserve is a rain shadow region. The Attappady reserve covers a total area of about 13,000 ha. A coffee and pepper estate is situated in this area along with semi-evergreen and moist deciduous forests.

The Silent Valley remained undisturbed due to its mountainous topography, which made accessibility difficult. The Attappady forest is subjected to heavy poaching pressure as it is close to the vested forests of Palghat. These two study sites are separated by about 20 km, but the levels of disturbance and vegetation types differ. There was also a difference of 400 m in elevation between the two sites.

History

The ownership of Silent Valley reserve forest was with the Government from very old days. Four hundred hectares of land near Walakkad were first given to private planters for planting from 1847 to 1873. When the planters abandoned the area, the Government took over the land in 1889. The human interference and collection of timber started as early as 1901. Human settlements were never established in the Silent Valley. In the nineteen seventies the Kerala Government proposed the building of a Hydro-Electric Project and the initial steps were taken. There was considerable influx of labourers for this work. This caused severe disturbance to the area. The Kerala Sasthra Sahithya Parishad, the Bombay Natural History Society and several other non-governmental organisations took up the cause of the conservation of the Silent Valley forests and pleaded for abandoning the Hydro-Electric Project. Eventually the Government abandoned the project and declared 90 km^2 as Silent Valley National Park. Management of the area was based on regular working plans from 1933-1934.

Geology

The rock formations of the study area are of Archean age and comprise gneisses and granulate of the khondalite snite of rocks intruded by later granites, pegmatites, and veinqartz. Rock types were garnetiferous - illuminate - biotite (muscovite) gneiss, biotite gneiss, cale granulite, amphibolite and granites. According to the Geological Survey of India, the Silent Valley area has rocks of well foliated gneisses to grantioid gneisses and granites referable to Archean Complex. According to Balagopalan (1990) the soils are rich in the clay fraction. With the increase of depth, clay contents increase indicating the infiltration of finer particles into deeper layers.

Climate

There are two distinct seasons in the study area. The monsoon season starting from end of May up to middle of November and the dry summer season from December to the first half of May. There is no clearly marked winter. During the monsoon days fast winds blow from the western side.

Silent Valley : Rainfall in each month at Silent Valley during the study period is given in Table 1.1.

Table 1.1: Rainfall data of Silent Valley

Months	Monthly rainfall in mm					
	1988	1989	1990	1991	1992	1993
January	-	Nil	Nil	Nil	1.6	Nil
February	-	Nil	Nil	Nil	Nil	202
March	-	Nil	6.2	11.4	Nil	14.4
April	-	72.6	13.4	117.7	120.8	6.2
May	-	70.2	803.1	30.8	249.7	-
June	-	888.7	836.4	1421.6	1532.7	-
July	1045.2	1642.0	954.8	1888.4	2043.8	-
August	1317.9	714.0	1069.2	593.4	1102.8	-
September	741.3	841.0	344.6	143.8	92.8	-
October	206.3	534.0	256.2	422.3	379.4	-
November	48.5	64.8	86.7	127.2	307.3	-
December	Nil	2.0	Nil	Nil	Nil	-

(-) no data available

Over the different seasons the temperature and relative humidity at Silent Valley did not show much variation (Table 1.2).

Table 1.2: Temperature and humidity at Silent Valley

Months	Mean temp. (°C)	Mean RH (%)
January	22	88
February	21	86
March	19	76
April	20	86
May	20	84
June	19	89
July	19	89
August	20	90
September	21	89
October	20	88
November	22	88
December	22	88

RH = Relative Humidity

Mukkali

Compared to Silent Valley, rainfall was less at Mukkali. This is evident from Table 1.3. Temperature was high at Mukkali and it varied from 21° C to 27° C in different months (Table 1.4).

Table 1.3: Rainfall data of Mukkali

Months	Monthly rainfall in mm					
	1988	1989	1990	1991	1992	1993
January	Nil	Nil	29.0	20.6	Nil	Nil
February	Nil	Nil	Nil	2.6	Nil	6.0
March	44.0	40.0	45.4	4.4	Nil	3.6
April	116.0	93.9	57.8	130.2	41.2	Nil
May	263.0	71.0	628.6	42.8	110.4	185.5
June	1055.0	928	579.6	1412.2	895.8	-
July	1107.0	1259.2	648.2	1623.4	1321.0	-
August	984.0	229.1	672.0	406.0	666.6	-
September	902.0	531.9	88.4	224.8	234.8	-
October	317.0	521.8	455.6	712.1	682.6	-
November	43.0	146.8	90.8	431.0	170.4	-
December	Nil	27.4	29.0	Nil	Nil	-

(-) no data available

Table 1.5: Temperature recorded at Mukkali

Months	1988	1989	1990	1991	1992	1993	Mean
January	-	21	21	22	24	25	23
February	-	21	23	23	26	26	24
March	-	23	24	26	22	27	24
April	-	27	27	27	27	27	27
May	-	26	25	30	26	26	27
June	24	25	23	23	27	-	24
July	23	24	23	22	24	-	23
August	23	22	22	23	23	-	23
September	24	23	24	25	25	-	24
October	22	23	25	25	25	-	24
November	23	23	23	25	25	-	24
December	22	22	22	23	26	-	23

(-) no data available

Vegetation

Silent Valley : The Western Ghats of Kerala has less than 8,000 km^2 of forest area in a highly fragmented form, which lies in six distinct tracks or clusters (Nair, 1988) and Silent Valley and Attappady lie in one of them. A total of 966 species of angiosperms, belonging to 559 genera and 134 families were recorded from Silent Valley and adjacent areas (Manilal, 1988).

According to Champion and Seth (1968) the forests found in Silent Valley were classified as Type 1A/C4 (vi) high-level evergreen forests of Palghat. Pascal (1988) described the vegetation of this area as *Cullenia exarillata-Mesua ferrea-Pallaquium ellipticum* type. It is characterised by the abundance of the three species and these may constitute about 80 per cent of the large trees. Top level is composed of *Cullenia exarillata, Machilus macrantha* Nees, *Elaeocarpus tuberculatus* Roxb., *Palaquium ellipticum* Engl., *Mesua ferrea* L., *Calophyllum elatum* Bedd., *Canarium strictum* Roxb., *Dysoxylum malabaricum* Bedd., *Vateria indica* L., *Poeciloneuron indicum* Bedd., *Eugenia* sp., *Mangifera indica* L., *Artocarpus heterophyllus* Lamk., *Cinnamomum zeylanicum* Bl., *Hopea glabra* W.& A.

Middle level by *Myristica laurifolia* Hk. f. & T., *Hydnocarpus laurifolia* Dennst., *Euphorbia longan* Lour., *Garcinia spicata* Hk. f., *Elaeocarpus serratus* L., *Gordonia obtusa* Wall., *Syzygium* sp., *Scolopia renata* Clos.,

Xanthophyllum flavescens Roxb., *Actinodaphne hookeri* Bedd., *Ochlandra beddomei* Gamb.

The lower level is composed of species like *Euonymus angulatus* W., *Agrostistachys indica* Dalz., *A. longifolia* Benth., *Syzygium laetum* (Buch. Ham.) *Gandhi, Paramignya beddomei* Tan., *Sauropus albicans* Bl., *Saprosma fragrans* Bedd., *Clerodendron viscosum* Vent., *Strobilanthes* sp., *Laportea crenulata* Gand., *Olea dioica* Roxb., *Linociera malabarica* Wall., *Pavetta zeylanica* Gamb.

Monocotyledons found in the area are *Arenga wighti* Griff., *Calamus* sp., *Pandanus furcatus* Roxb. etc. Ground flora include *Elettaria cardamomum* Mant., *Asparagus racemosus* Willd., *Curcuma* sp., *Heckeria subpeltata* Kunth.

Climbers such as *Gnetum scandens* Roxb., *Calamus* sp., *Smilax zeylanica* L., *Derris* sp., *Entada pursaetha* DC., and *Morinda* sp. also were seen.

Angiosperm flora of the Silent Valley consists of species belonging to 134 families and 53 genera. Among them there are 701 Dicotyledons from 113 families and 420 genera, and 265 monocotyledons from 21 families and 139 genera. The dominant plant families recorded from the Silent Valley are Orchidaceae, with 108 species from 49 genera, Poaceae with 56 species belonging to 32 genera; Fabaceae with 55 species representing 26 genera; Rubiaceae with 49 species of 27 genera and Asteraceae with 45 species representing 25 genera (Manilal, 1988).

Other types of vegetation

Other type of vegetation of the area is Southern Secondary Moist Mixed Deciduous Forest. The major species are *Xylia xylocarpa* Taub., *Terminalia paniculata* Roth, *Grewia tiliaefolia* Vahl, *Lagerstroemia lanceolata* Wall., *Dalbergia latifolia* Roxb., *Albizzia odoratissima* Benth., *Bombax malabarica* DC., *Pterocarpus marsupium* Roxb., *Terminalia tomentosa* W. & A., *Careya arborea* Roxb., *Randia dumetorum* Lam., Shrub layer consists of *Helicteres isora* L., and *Clerodendron infortunatum* L. Bamboos are also common. Trees on the top of the hills are of stunted growth not attaining much height.

Grasslands

In grasslands, near the tropical forests at Silent Valley, *Phyllanthus* sp., *Dalbergia* sp. and *Careya* sp. are common. *Phoenix* is seen at some places.

Kole wetlands

Location and topography

The Kole wetlands lie between 10° 20' and 10° 40' N latitudes and 75° 58' and between 76° 11' E longitudes (Fig. 1.2). The Kole wetlands with an extent of 13,632 ha are spread over Thrissur and Malappuram Districts in Kerala State. It extends from the northern banks of Chalakudy River in the South to the southern banks of Bharathapuzha River in the North. Eastern side of Kole wetlands is Thrissur town and western side extends up to Arabian Sea. The name "Kole" refers to the peculiar type of paddy cultivation carried out from December to May and this Malayalam word indicates bumper yield of high returns in case floods do not damage the crops.

History

The Kole wetlands were reclaimed from Thrissur *Kayal* for rice cultivation at the beginning of the 18th century. Erstwhile Maharaja permitted to convert these wetlands into paddy fields (Johnkutty and Venugopal, 1993). The Kole wetland, which forms the rice granary of Thrissur and Malappuram Districts, comprises a unique ecosystem in Kerala. The water pumped out from the fields was stored in a network of canals interspersed throughout the area. The water and the timely showers used to produce bumper crops, if other conditions were favourable.

Physiography

Physiographically, the area is unique and the entire tract is a product of fluvial estuarine agencies modified by human activities. The area is devoid of any significant relief features and consists of extensive flat land surface interspersed with uplands. The area is saucer shaped with low lands at the centre with elevation gradually increasing towards the fringes. The land around the rice fields have steep slopes which are terraced and put under perennial crops like arecanut and coconut and annual crop like banana and yams. The slopes merge with level plateau lands. The dry lands of the Kole region adjoining the coastal belt have level topography and are under coconut cultivation. The Kole wetlands are low lying tracts located 0.5 to 1 m below MSL and it remains submerged for about six months in a year. These lands were formerly shallow lagoons, which gradually were silted up.

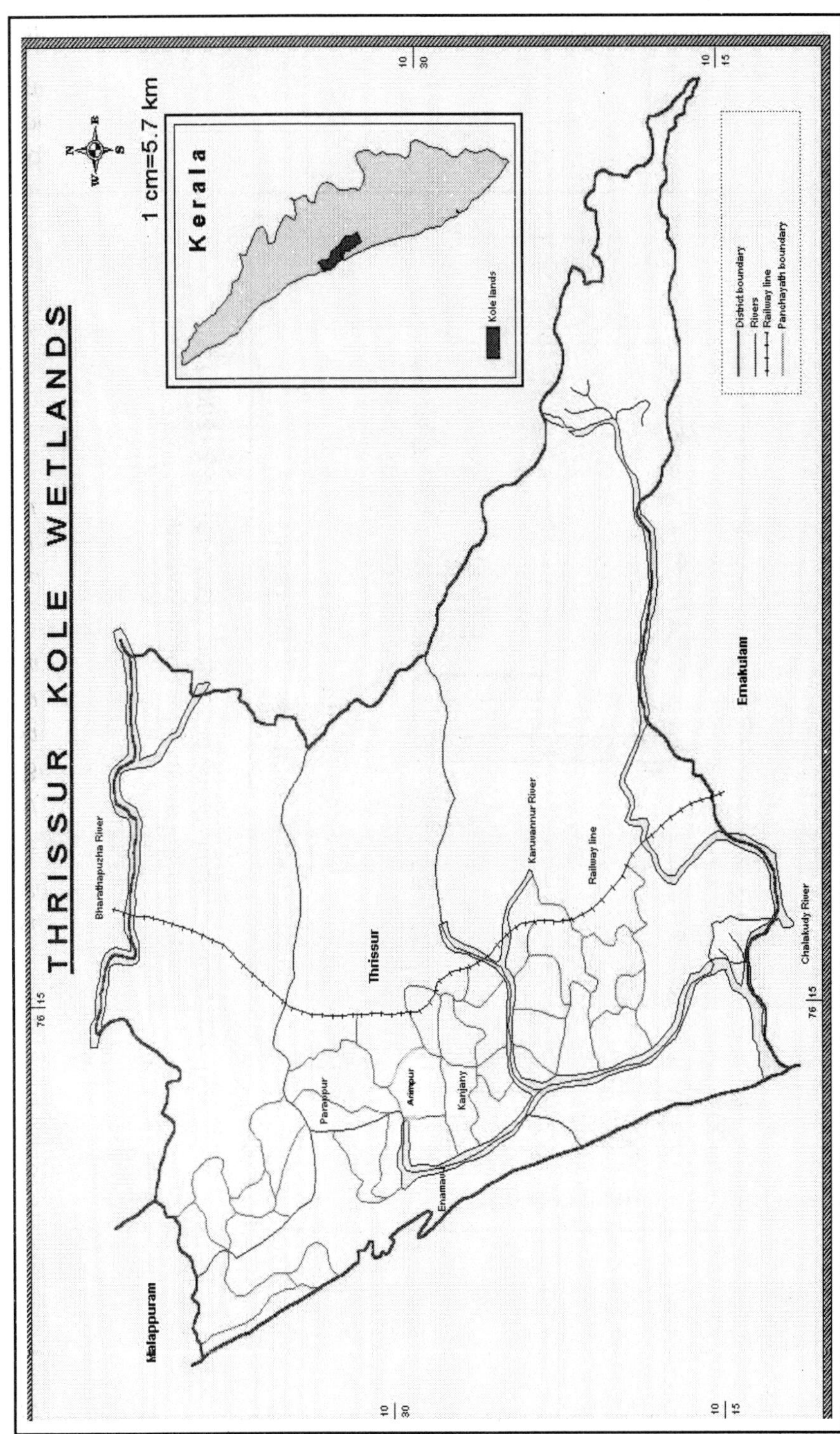

Fig. 1.2: Kole wetland of Kerala.

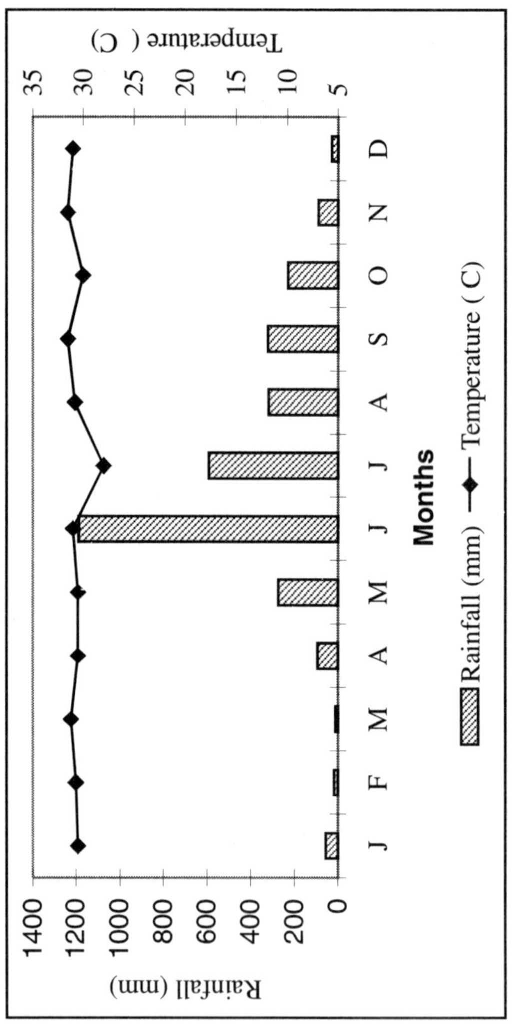

Fig. 1.3: Ombro-thermic graph of the Kole wetlands (1992 - 2001)
(Source: Irrigation Department, Thrissur)

Rivers, watercourses and drainage

Keecheri and Karuvannur Rivers bring the floodwater into the wetlands, which finally empty into the Arabian Sea. These rivers in spate discharge the flood waters into the low lying Kole wetlands and raise water depth to more than 550 cm. The Kole wetlands function as the flood basin for both the rivers. The Karuvannur River has two tributaries *viz.* Manali and Kurumali. Kurumali is formed of two tributaries, Chimmoni and Mupli. All these streams start from the Western Ghats and flow along steep slopes until they reach the plains where they take very meandering courses and join to form the main river in the plains. When it reaches the West, the river branches into two, one going directly to North joins the Chettuva Lake and the other flowing South joins the Manakodi Lake. The Keecheri River flows down from Machad hills, traverses West, then turns South, and joins the Kole wetlands on the northern side draining finally into Enamakkal Lake, which is connected to Chettuva Lake. The river though small, has flash floods during monsoon.

Climate and season

The climate of the area is moderate and there are three different distinct seasons. The dry season (December to April), wet season-I (May to August) during the period of southwest monsoon and wet season-II (September to November) during northeast monsoon. About 60 per cent of the rainfall is obtained during the southwest monsoon, 30 per cent during the northeast monsoon and the remaining 10 per cent in summer. The average annual rainfall is 3,200 mm (James, 1983) and there is a variation in the temporal distribution of rainfall (Fig. 1.3). The maximum rainfall is received during the month of June followed by July. Extremes of heat or cold are not felt and the temperature varied from 28° C to 31.5° C. Atmosphere is always damp along the coastal belt due to high humidity.

❑❑❑

Chapter 2

Species Composition of Birds

Species composition of birds in an area is related to the type of vegetation, height above the sea level, the availability of microhabitats and various other factors. The study site at Silent Valley covered all the representative habitats namely forest canopy, forest interior, forest edge, rocky patches, grass patches and foliage air interface met within a tropical forest. Riparian patches were excluded after an initial survey as it contained only a few species of forest birds. The Mukkali study site covered moist deciduous forest, burned forests, coffee estate, open rocky areas and agricultural land. Common species, Dominant species and Species richness of the two study areas were assessed and the data are presented in this chapter.

METHODS

Species richness and species composition of birds in the study sites were computed from the census data and from field observations. Only those species of birds, which were recorded from the intensive study areas were included in the analysis. The number of species recorded is considered as species richness. The collector's curve method is used for assessing sampling efficiency. Collector's curve was drawn by plotting the cumulative number of species observed in each month against the months of observation (Pielou, 1975).

To find out the common bird species of each area Commonness Index of the two areas was computed. Commonness Index is the average sighting frequency of a species in one sampling. The relative dominance of each bird species in the two areas was determined by calculating Dominance Index. The following formula is used for calculating Relative Dominance.

$$\text{Relative Dominance} = \frac{n_i}{N} \times 100$$

where

n_i = number of individuals in the species.

N = The total number of individuals of all the species seen during the study period

Collector's curve

The sampling efficiency is determined by studying this curve. In vegetational studies this curve is generally known as species-area curve. Collector's curves for Silent Valley and Mukkali are shown in Fig. 2.1. An increase in the occurrence of species is observed up to October 1992. After this, the rise in the curve slopes down which indicates that the chances of encountering new species appearing in the observations are very few, this shows that the area is adequately covered. Curves for both the areas show the identical trend. Further observations from this point onwards may not be worthwhile in enumerating new species.

Occurrence of species

Silent Valley

Ninety-nine taxa of birds, belongs to 10 Orders and 31 Families were recorded from the Silent Valley. Presence of species in different months over the years was presented in Table 2.1. The most common species found at Silent Valley was Yellow-browed Bulbul followed by White-cheeked Barbet, Bush Chat and Common Hill Myna. Commonness Index of 10 most common species at Silent Valley is given in Table 2.2.

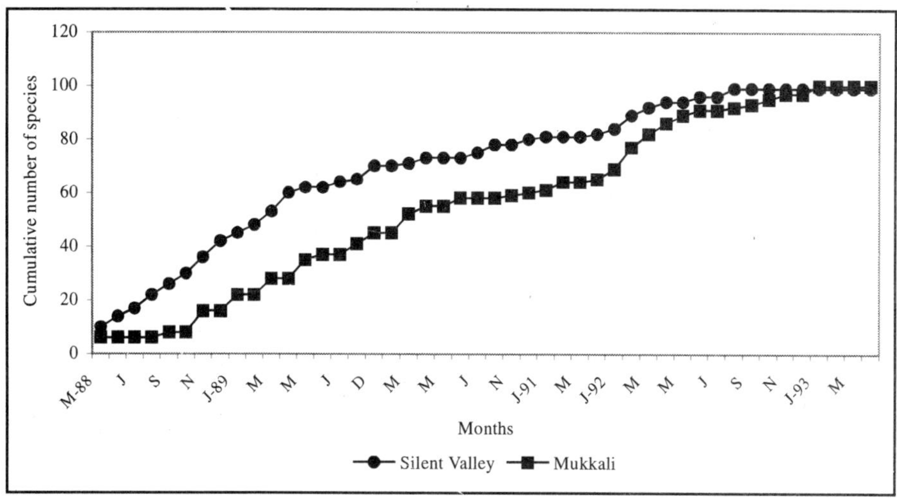

Fig. 2.1: Collector's curves from the study area

Table 2.1: Occurrence of birds at Silent Valley in different months

S. No.	Species	Months J	F	M	A	M	J	J	A	S	O	N	D
1.	Indian Pond-Heron	√	-	√	-	√	√	-	-	-	-	-	-
2.	Black-shouldered Kite	√	√	√	-	-	-	-	√	-	-	-	-
3.	Brahminy Kite	-	-	√	-	-	-	-	-	-	√	-	-
4.	Shikra	-	-	√	-	-	-	√	-	-	-	√	-
5.	Black Eagle	√	-	-	√	√	√	-	-	√	-	√	√
6.	Crested Serpent-Eagle	-	-	√	-	√	-	-	√	-	√	√	-
7.	Peregrine Falcon	-	-	-	-	-	-	-	-	-	-	√	√
8.	Painted Bush-Quail	√	√	√	√	√	√	√	√	-	-	-	√
9.	Red Spurfowl	-	-	√	-	-	-	-	√	-	-	√	-
10.	Grey Junglefowl	√	√	√	√	√	√	√	-	√	√	√	√
11.	Greyfronted Green Pigeon	√	-	√	√	√	-	-	-	-	-	-	-
12.	Yellow-legged Green-Pigeon	√	-	√	-	√	-	-	-	-	-	√	√
13.	Pombadour Green-Pigeon	√	-	-	-	-	-	-	-	-	-	-	
14.	Mountain Imperial-Pigeon	√	√	√	√	-	-	-	√	-	-	-	-
15.	Blue Rock Pigeon	-	-	√	-	-	-	-	-	-	-	-	-
16.	Nilgiri Wood-Pigeon	√	-	-	-	-	-	-	-	-	-	-	-
17.	Spotted Dove	√	-	-	-	-	-	-	-	-	-	-	√
18.	Emerald Dove	√	-	√	√	-	-	-	-	-	-	√	√

Contd...

19.	Rose-ringed Parakeet	√	-	√	-	√	√	-	√	√	√	√	√
20.	Blossom-headed Parakeet	√	√	√	√	√	√	√	√	√	√	√	√
21.	Blue-winged Parakeet	√	-	√	√	-	-	√	√	-	-	√	√
22.	Indian Hanging-Parrot	√	-	√	-	-	-	-	√	-	√	√	√
23.	Red-winged Crested Cuckoo	-	-	-	-	-	-	-	√	-	-	-	-
24.	Greater Pheasant	-	-	√	-	-	-	-	-	-	-	-	-
25.	Short-eared Owl	√	-	√	-	-	-	-	-	-	-	-	-
26.	White-rumped Needletail-Swift	-	√	-	-	-	-	-	√	-	-	-	√
27.	Malabar Trogon	-	-	√	-	-	-	-	-	-	-	-	-
28.	Chestnut-headed Bee-eater	√	-	-	√	-	-	-	√	-	-	-	√
29.	Malabar Grey Hornbill	√	√	√	-	-	√	√	-	√	-	-	√
30.	Great Pied Hornbill	-	√	-	-	-	√	-	-	-	-	-	-
31.	White-cheeked Barbet	√	√	√	√	√	√	-	√	√	√	√	√
32.	Speckled Piculet	-	-	-	-	-	-	-	√	√	-	-	-
33.	Lesser Golden-backed Woodpecker	√	√	√	√	√	√	√	√	√	√	√	√
34.	Common Golden-backed Woodpecker	√	√	-	√	-	-	-	-	-	-	-	-
35.	Great Black Woodpecker	√	-	-	√	√	-	-	-	-	-	-	-
36.	Heart-spotted Woodpecker	-	-	-	√	√	√	-	-	-	-	-	-
37.	Indian Pitta	√	√	-	-	-	-	-	-	-	-	-	√
38.	House Swallow	√	-	√	√	-	-	-	-	-	-	√	√
39.	Red-rumped Swallow	√	√	√	√	√	√		√	√	√	√	√
40.	Rufous-backed Shrike	√	-	-	-	-	-	-		-	-	-	-
41.	Eurasian Golden Oriole	√	-	-	-	-	-	-	-	-	-	-	-
42.	Black-naped Oriole	√	-	-	-	-	-	-	-	-	-	-	√
43.	Black-headed Oriole	√	-	√	-	-	-	-	-	-	-	-	-
44.	Black Drongo	√	√	√	√	√	-	-	√	-	√	√	√
45.	White-bellied Drongo	-	-	-	-	-	-	-	-	-	-	√	√
46.	Bronzed Drongo	√	-	√	-	-	-	-	-	-	-	√	√
47.	Greater Racket-tailed Drongo	√	√		√	√	√	√	√	√	-	√	√
48.	Common Myna	-	-	-	-	√	-	-	-	-	-	-	-
49.	Common Hill-Myna	√	√	√	√	√	√	-	-	√	-	√	√
50.	Indian Treepie	-	-	-	-	-	-	-	-	√	-	-	-
51.	White-bellied Tree Pie	√	√	√	√	√	√	√	√	√	√	√	√
52.	Jungle Crow	-	-	√	-	-	-	-	-	-	-	-	-
53.	Scarlet Minivet	√	-	√	√	√	√	√	√	√	√	√	√
54.	Common Iora	-	-	-	√	-	-	-	-	-	-	-	-
55.	Gold-fronted Chloropsis	√	-	-	√	-	-	-	-	-	-	-	-
56.	Asian Fairy-Bluebird	-	-	-	√	-	-	-	-	-	-	-	-
57.	Black-crested Bulbul	-	-	-	-	-	-	-	-	-	-	√	-
58.	Red-whiskered Bulbul	√	√	√	√	√	√	√	√	√	√	-	√

Contd...

59.	Red-vented Bulbul	-	-	√	-	-	-	√	√	-	-	√	√
60.	Yellow-browed Bulbul	√	√	√	√	√	√	√	√	√	√	√	√
61.	Black Bulbul	√	√	√	√	-	-	-	-	√	-	√	√
62.	Spotted Babbler	-	-	√	-	-	-	√	-	-	-	-	-
63.	Hodgson's Scimitar-Babbler	√	-	-	√	√	-	-	-	-	-	-	-
64.	Black-headed Babbler	-	-	-	-	-	-	-	√	√	-	-	-
65.	Indian Rufous Babbler	-	-	-	-	-	-	-	-	-	-	-	√
66.	Jungle Babbler	√	√	√	√	√	√	√	√	√	√	√	√
67.	White-headed Babbler	-	√	-	-	-	-	-	-	-	-	-	-
68.	Asian Brown Flycatcher	-	-	√	-	-	-	-	-	-	-	-	-
69.	Brown-breasted Flycatcher	-	-	-	-	-	-	√	-	-	-	-	-
70.	White-bellied Blue-Flycatcher	-	-	-	-	-	-	-	-	-	-	√	-
71.	Tickell's Blue-Flycatcher	√	√	√	-	√	-	-	-	-	-	-	.√
72.	Verditer Flycatcher	-	-	-	-	-	-	-	-	-	-	-	√
73.	Asian Paradise-Flycatcher	√	-	-	-	-	-	-	-	-	-	√	-
74.	Greenish Leaf-Warbler	√	√	-	-	√	√	-	-	√	-	√	√
75.	Oriental Magpie-robin	-	-	-	-	√	-	-	-	-	-	-	-
76.	Pied Bushchat	√	√	√	√	√	√	√	√	√	-	√	√
77.	Malabar Whistling-Thrush	√	√	√	√	√	√	√	√	√	√	-	√
78.	Orange-headed Thrush	-	-	-	-	-	-	-	-	-	-	-	√
79.	White-throated Ground Thrush	√	√	-	-	-	-	-	-	-	-	√	√
80.	Eurasian Blackbird	√	-	-	√	-	√	-	-	-	-	√	-
81.	Great Tit	-	-	-	-	√	-	-	-	-	-	-	-
82.	Black-lored Tit	-	-	√	√	√	√	√	-	-	-	√	-
83.	Velvet-fronted Nuthatch	-	-	-	√	√	√	-	-	-	-	-	-
84.	Oriental Tree Pipit	√	-	-	-	-	-	-	-	-	-	-	-
85.	Forest Wagtail	-	-	√	-	-	-	-	-	-	-	-	-
86.	Yellow Wagtail	√	√	√	-	-	-	-	-	√	√	√	√
87.	Grey Wagtail	-	-	-	-	-	-	-	-	-	-	√	-
88.	Thick-billed Flowerpecker	-	-	-	-	-	-	-	-	√	-	-	-
89.	Tickell's Flowerpecker	√	-	√	-	√	√	-	√	√	-	√	-
90.	Purple-rumped Sunbird	-	-	-	-	-	-	-	-	√	-	-	-
91.	Small Sunbird	√	√	√	√	-	√	√	√	√	√	√	√
92.	Loten's Sunbird	-	-	-	-	-	-	-	-	-	√	√	-
93.	Little Spiderhunter	-	√	√	√	√	-	-	-	-	-	-	-
94.	Nilgiri White-eye	-	-	-	√	√	√	-	√	√	-	√	-
95.	White-throated Munia	√	-	-	-	-	-	-	-	-	-	-	-
96.	Rufous-bellied Munia	-	-	-	-	-	-	-	-	√	-	-	-
97.	Spotted Munia	-	-	-	-	-	-	-	-	-	-	-	√
98.	Black-headed Munia	√	-	-	-	-	-	-	√	-	-	-	√
99.	Common Rosefinch	√	√	√	-	-	-	-	-	-	-	-	√

√ = Present; - = Not recorded

Table 2.2: Commonness index of selected species at Silent Valley

S. No.	Species	Commonness Index
1.	Yellow-browed Bulbul	6.29
2.	White-cheeked Barbet	1.78
3.	Pied Bush Chat	1.75
4.	Common Hill Myna	1.44
5.	Black Bulbul	1.32
6.	Red-whiskered Bulbul	1.18
7.	Malabar Whistling Thrush	0.98
8.	Grey Junglefowl	0.96
9.	Southern Tree Pie	0.83
10.	Lesser Golden-backed Woodpecker	0.81

The dominant species in the community at Silent Valley were, the Yellow-browed Bulbul, the Black Bulbul, the Common Hill Myna, the Jungle Babbler and the Pied Bush Chat. Dominance index of most dominant 10 species at Silent Valleys in given in Table 2.3.

Table 2.3: Dominance index of selected species at Silent Valley

S. No.	Species	Dominance Index
1.	Yellow-browed Bulbul	20.33
2.	Black Bulbul	9.27
3.	Common Hill Myna	8.04
4.	Jungle Babbler	4.56
5.	Pied Bush Chat	4.29
6.	White-cheeked Barbet	4.19
7.	House Swallow	3.42
8.	Red-whiskered Bulbul	3.38
9.	Rose-ringed Parakeet	2.79
10.	Black Drongo	2.43

Mukkali

Ninety-six taxa of birds belongs to 10 Orders and 30 Families were recorded from Mukkali study site. Presence of species in different months over the years is presented in Table 2.4.

Table 2.4: Occurrence of birds at Mukkali in different months

S. No.	Species	Months											
		J	F	M	A	M	J	J	A	S	O	N	D
1.	Black-shouldered Kite	√	-	-	√	-	√	√	-	√	√	√	√
2.	Shikra	-	-	-	√	-	-	-	-	-	-	-	-
3.	Black Eagle	-	-	√	-	-	-	-	-	-	√	-	-
4.	Crested Serpent-Eagle	-	-	-	-	-	√	-	-	-	-	-	√
5.	Peregrine Falcon	-	-	-	-	-	-	-	-	√	-	-	-
6.	Grey Jungle Fowl	√	√	√	√	√	-	-	-	√	√	√	√
7.	Pombadour Green-Pigeon	-	-	-	-	√	-	-	-	-	-	-	-
8.	Spotted Dove	√	√	√	√	√	√	√	√	√	√	√	√
9.	Rose-ringed Parakeet	√	√	√	√	√	√	√	-	√	√	√	√
10.	Blossom-headed Parakeet	√	√	√	√	-	-	-	-	√	√	√	√
11.	Blue-winged Parakeet	-	√	√	√	-	-	-	-	-	-	-	-
12.	Indian Hanging-Parrot	-	√	√	√	√	√	-	-	-	-	-	-
13.	Brainfever Bird	-	-	-	-	√	-	-	-	-	-	-	-
14.	Asian Koel	-	-	√	-	-	-	-	-	-	-	-	-
15.	Greater Pheasant	-	-	-	-	√	-	-	-	-	-	-	√
16.	Jungle Owlet	√	-	-	√	√	-	-	-	-	-	-	-
17.	Short-eared Owl	-	√	-	-	-	-	-	-	-	-	-	-
18.	Malabar Trogon	-	-	√	-	-	-	-	-	√	-	√	-
19.	White-breasted Kingfisher	-	-	√	√	-	√	√	-	-	-	-	√
20.	Chestnut-headed Bee-eater	-	-	-	√	√	-	-	-	-	-	√	√
21.	Common Hoopoe	-	-	√	-	-	√	-	-	-	-	-	-
22.	Malabar Grey Hornbill	-	-	-	√	-	√	√	-	-	-	-	√
23.	Great Pied Hornbill	-	-	-	-	-	√	-	-	-	-	-	-
24.	White-cheeked Barbet	√	√	√	√	√	√	√	√	√	√	√	√
25.	Small Yellow-naped Woodpecker	-	-	-	-	-	√	-	-	-	-	-	-
26.	Common Golden-backed Woodpecker	-	-	√	-	-	-	-	√	√	-	-	√
27.	Lesser Golden-backed Woodpecker	√	√	√	√	√	√	√	√	√	√	√	√
28.	Heart-spotted Woodpecker	-	-	√	√	-	-	-	-	√	-	-	√
29.	Indian Pitta	√	-	√	√	-	-	-	-	-	-	-	-
30.	House Swallow	√	√	√	-	√	-	-	√	-	-	√	√
31.	Red-rumped Swallow	-	-	-	-	-	√	-	-	-	-	-	-
32.	Greater Grey Shrike	-	-	-	√	-	-	-	-	-	-	-	-
33.	Bay-backed Shrike	√	-	-	√	-	√	√	-	-	-	-	-
34.	Rufous-backed Shrike	√	√	√	-	-	-	-	-	-	-	-	-
35.	Eurasian Golden Oriole	-	√	-	-	-	-	-	-	-	-	-	-

Contd...

36.	Black-headed Oriole	-	√	√	√	√	√	-	-	√	-	√	-
37.	Black Drongo	√	√	√	√	√	√	√	-	√	√	√	√
38.	Bronzed Drongo	√	√	-	√	-	√	√	√	√	√	√	-
39.	Spangled Drongo	-	-	-	-	-	√	-	-	-	-	-	-
40.	Greater Racket-tailed Drongo	√	√	√	√	√	√	√	√	√	√	√	√
41.	Common Myna	√	√	√	√	√	√	-	-	-	-	-	-
42.	Jungle Myna	√	-	√	√	-	√	-	-	-	-	-	-
43.	Common Hill-Myna	√	√	-	-	√	√	-	-	-	-	-	-
44.	Indian Treepie	√	-	√	√	√	-	√	√	√	√	√	-
45.	White-bellied Tree Pie	√	-	√	-	√	√	√	-	√	-	√	-
46.	House Crow	-	√	-	-	√	-	√	-	√	-	√	-
47.	Jungle Crow	√	-	√	-	-	-	-	-	-	-	-	-
48.	Cuckoo Shrike	-	-	√	-	-	-	-	-	√	-	-	-
49.	Scarlet Minivet	√	-	√	√	-	√	√	-	√	√	√	√
50.	Common Iora	-	√	√	√	-	-	√	-	-	-	-	-
51.	Gold-fronted Chloropsis	√	-	√	√	-	√	√	-	√	-	√	√
52.	Jerdon's Chloropsis	-	-	√	-	-	√	-	-	-	√	-	-
53.	Asian Fairy-Bluebird	-	-	-	√	-	√	-	-	-	-	-	-
54.	Black-crested Bulbul	-	-	√	√	-	-	-	-	√	-	-	-
55.	Red-whiskered Bulbul	√	√	√	√	√	√	√	√	√	√	√	√
56.	Red-vented Bulbul	√	√	√	√	√	√	√	√	√	√	√	√
57.	Yellow-browed Bulbul	√	√	√	√	-	√	√	√	√	√	√	√
58.	Black Bulbul	√	-	√	-	-	√	-	-	√	√	√	√
59.	Hodgson's Scimitar-Babbler	-	-	-	-	-	-	-	-	-	√	-	-
60.	Indian Rufous Babbler	√	-	√	-	-	-	-	-	-	√	-	-
61.	Jungle Babbler	√	√	√	√	√	√	√	√	√	√	√	√
62.	White-headed Babbler	√	-	√	√	√	√	-	-	-	√	√	-
63.	Asian Brown Flycatcher	√	-	√	-	-	-	-	√	-	-	√	-
64.	Brown-breasted Flycatcher	√	-	-	-	-	-	-	-	-	-	-	-
65.	Rusty-tailed Flycatcher	√	-	-	-	-	-	-	-	-	-	-	-
66.	Black-and-Orange Flycatcher	-	√	-	-	-	-	-	-	-	-	-	-
67.	Verditer Flycatcher	-	√	-	-	-	-	-	-	-	-	-	-
68.	Asian Paradise-Flycatcher	-	-	√	-	-	-	-	-	-	-	√	-
69.	Common Tailorbird	-	-	-	√	√	-	-	-	-	√	√	-
70.	Greenish Leaf-Warbler	√	√	-	-	√	-	-	-	-	√	√	-
71.	Oriental Magpie-robin	√	√	√	√	√	√	√	-	√	-	√	√
72.	Pied Bushchat	√	√	√	-	-	√	-	√	√	√	√	√
73.	Blue Rock-Thrush	√	√	√	-	-	-	-	-	√	-	-	-
74.	Malabar Whistling-Thrush	-	-	√	√	-	√	√	√	-	-	√	-
75.	Orange-headed Thrush	√	-	-	-	-	-	-	-	-	-	-	-

Contd...

76. Whitethroated Ground Thrush	√	√	√	√	-	√	-	-	-	-	√	√
77. Eurasian Blackbird	√	√	-	-	-	-	-	-	-	-	-	-
78. Great Tit	-	√	√	-	-	-	-	-	-	-	-	-
79. Black-lored Tit	-	√	√	-	-	-	√	-	-	-	-	-
80. Velvet-fronted Nuthatch	-	-	-	-	√	√	-	-	-	√	-	-
81. Forest Wagtail	√	-	√	-	-	-	-	-	-	-	-	-
82. Yellow Wagtail	√	√	√	-	-	-	-	-	√	√	√	√
83. Pied Wagtail	-	-	-	-	-	-	-	-	-	-	√	-
84. Tickell's Flowerpecker	√	√	-	-	-	-	-	-	-	-	-	√
85. Purple-rumped Sunbird	-	√	-	-	√	-	-	-	√	√	√	√
86. Small Sunbird	√	-	√	-	-	-	-	-	-	-	√	-
87. Loten's Sunbird	-	√	-	√	-	-	-	-	√	-	-	-
88. Little Spiderhunter	-	-	-	-	-	-	-	-	-	-	√	-
89. Oriental White-eye	-	√	-	-	-	√	-	-	-	-	√	-
90. Yellow-throated Sparrow	-	√	-	-	-	-	-	-	-	-	-	-
91. Rufous-bellied Munia	-	-	-	-	-	√	-	-	-	-	-	-
92. Black-headed Munia	√	-	-	-	-	-	√	-	√	√	-	-

√ = Present; - Not recorded

The most common species found at Mukkali were Black Drongo, White-cheeked Barbet, Jungle Babbler, Red-vented Bulbul and Racket-tailed Drongo. Commonness index of birds with highest values is presented in Table 2.5. The most dominant species found in the area were determined from the dominance index. Jungle Babbler, Red-Whiskered Bulbul and Black Drongo were the most dominant species found at Mukkali. Dominance index of 10 species are listed in Table 2.6.

Table 2.5: Commonness index of selected species at Mukkali

S. No.	Species	Commonness Index
1.	Black Drongo	2.08
2.	White-cheeked Barbet	2.04
3.	Jungle Babbler	1.74
4.	Red-vented Bulbul	1.47
5.	Racket-tailed Drongo	1.17
6.	Yellow-browed Bulbul	0.89
7.	Spotted Dove	0.85
8.	Lesser Golden-backed Woodpecker	0.83
9.	Oriental Magpie Robin	0.70
10.	Blossom-headed Parakeet	0.68

Table 2.6: Dominance index of selected species at Mukkali

S. No.	Species	Dominance Index
1.	Jungle Babbler	19.08
2.	Red-whiskered Bulbul	6.19
3.	Black Drongo	5.94
4.	White-cheeked Barbet	5.47
5.	Red-vented bulbul	4.61
6.	Blossom-headed Parakeet	3.42
7.	Racket-tailed Drongo	3.17
8.	Yellow-browed Bulbul	3.13
9.	Scarlet Minivet	2.63
10.	Rose-ringed Parakeet	2.41

Unique species

Thirty species listed below was recorded only from Silent Valley.

S. No.	Species	S. No.	Species
1.	Indian Pond Heron	16.	White-bellied Drongo
2.	Brahminy Kite	17.	Spotted Babbler
3.	Shikra	18.	Black-headed Babbler
4.	Painted Bush Quail	19.	Brown Flycatcher
5.	Red Spurfowl	20.	White-bellied Blue Flycatcher
6.	Yellow-legged Green Pigeon	21.	Tickell's Blue Flycatcher
7.	Imperial Pigeon	22.	Nilgiri Flycatcher
8.	Blue-rock Pigeon	23.	Leaf Warbler
9.	Nilgiri Wood Pigeon	24.	Tree Pipit
10.	Emerald Dove	25.	Grey Wagtail
11.	Red-winged Crested Cuckoo	26.	Thick-billed Flowerpecker
12.	White-rumped Spinetail	27.	Tickell's Flowerpecker
13.	Speckled Piculet	28.	White-throated Munia
14.	Greatblack Woodpecker	29.	Spotted Munia
15.	Black-naped Oriole	30.	Rose Finch

Among this Imperial Pigeon and Red-winged Crested Cuckoo are rare birds. In the same way only 21 species were recorded from Mukkali, which are listed below:

S. No.	Species	S. No.	Species
1.	Brainfever Bird	11.	Small Grey Cuckoo Shrike
2.	Koel	12.	Goldmantled Chloropsis
3.	Jungle Owlet	13.	Rufous-tailed Flycatcher
4.	White-breasted Kingfisher	14.	Black-and-orange Flycatcher
5.	Hoopoe	15.	Verditer Flycatcher
6.	Small Yellownaped Woodpecker	16.	Common Tailor Bird
7.	Grey Shrike	17.	Blue Rock Thrush
8.	Bay-backed Shrike	18.	Pied Wagtail
9.	Rufous-backed Shrike	19.	Yellow-throated Sparrow
10.	Jungle Myna	20.	Haircrested Drongo
		21.	House Crow

Changes in bird species richness

Distinct changes in species composition were recorded among the birds of the Silent Valley and Mukkali over different months. During monsoon months the number of species present in the Silent Valley study site was low. But as the rain stopped, new species started to arrive and a maximum of fifty five species were recorded in the month of January (Table 2.7). In the monsoon season reduction in species richness was observed in all years. Similarly an increase of species richness was observed during summer in all the years. From this it is obvious that the bird species show a distinct change during different seasons. Mukkali data also shows the same trends as that of Silent Valley (Table 2.7).

Table 2.7: Monthly variation in the species richness at Silent Valley and Mukkali during the study period

Name of the area	J	F	M	A	M	J	J	A	S	O	N	D
Silent Valley	55	42	46	34	40	23	26	27	34	21	39	42
Mukkali	46	36	53	44	29	43	25	17	33	28	38	27

There is no significant difference in bird species richness between years in monsoon season and in summer season at Silent Valley (Table 2.8). But a significant difference is obtained between years in both seasons at Mukkali (Table 2.9). Less number of species recorded during the initial years of the study has contributed to the significant difference obtained in the number of species between years in both seasons at Mukkali.

Table 2.8: Seasonal variation in the bird species richness between years at the Silent Valley

Year	Monsoon	Summer
1988	35	—
1989	25	58
1990	42	49
1991	—	63
1992	35	55
1993	—	35
c^2	4.28	8.92
P < = 0.05	NS	NS

- not included in the analysis

Table 2.9: Seasonal variation in the bird species richness between years at Mukkali

Year	Monsoon	Summer
1988	17	—
1989	22	30
1990	30	39
1991	—	40
1992	63	64
1993	—	44
c^2	38.97	14.64
P < =	0.001	0.001

— not included in the analysis

Endemic and threatened species

The following species endemic to Western Ghats were recorded from the study area.

1. Nilgiri Wood Pigeon *Columba elphinstonii*
2. Blue-winged Parakeet *Psittacula columboides*

3. Malabar Grey Hornbill *Tockus griseus*
4. Southern Tree Pie *Dendrocitta leucogastra*

Among these the Nilgiri Wood Pigeon is a globally threatened species found in India. During the study, the birds were observed 4,500 times and a total of 9,921 birds were counted. Out of the 137 species located in the whole study area 21 migrant species were recorded from Silent Valley and 11 species were observed at Mukkali; others were residents. Fifty six species were common in both sites, while 30 species were found only at Silent Valley and 21 species seen only at Mukkali. Silent Valley is not a major wintering area of Palaearctic migrants and most of the birds were showing only local movements. The migrants which were recorded from here were the Wagtails (*Motacilla* sp.), Blackheaded Oriole (*Oriolus chinensis*), Rose Finch (*Carpodacus erythrinus*) and Redwinged Crested Cuckoo (*Calamator coromandus*). No wintering waterfowls were recorded from the area. This was due to the lack of wetlands in the study area. Most of the Doves, Pigeons, Parakeet and Black Bulbul (*Hypsipetes madagascariensis*) were not recorded during rainy season, but were seen returning to the area with the retreat of the rain.

The Yellow-browed Bulbul is the most common and dominant species at Silent Valley. The second common species, the Small Green Barbet, comes only in the sixth position in dominance. From the dominance index it is clear that barring a few species all other species are very rare. Red-whiskered Bulbul is the common and dominant species at Silent Valley than the Red-vented Bulbul. Similar observations were recorded by Vijayagopal (1991) also. At Mukkali most common species was the Black Drongo, but the most dominant species was the Jungle Babbler. Here also except a few species all the others were rare.

Due to the heavy mist and low activity of birds during monsoon, their detectability was poor and this may be one reason for the lower number of species recorded during monsoon days. In addition to this, local movement of species like Black Bulbul to the study area is observed during summer. It seems that rare species like Great Indian Hornbill and Great Black Woodpecker were affected severely during the pre-survey period of the abandoned hydroelectric project. Because most of the dry trees were burned for firewood and the Great Indian Hornbill was hunted for its flesh (Vijayan and Balakrishnan, 1977).

At present the population of these species is very low. Those species of birds recorded from Mukkali are found in the moist deciduous forests in other parts of the State also (Jayson and Nair, 1993).

❑❑❑

Chapter 3

Species Abundance Relations

In this chapter the data collected in Silent Valley and Mukkali on the abundance of birds are discussed. Species abundance relations of the community is also prepared and presented in this chapter.

METHODS

Abundance : Total number of birds seen in each month and seasons (summer and monsoon) were calculated using the census data.

Density : Density of birds in each month, in two seasons and individual density of selected species were also calculated. The density was estimated using the line transect method (Burnham *et al.*, 1980). The Fourier series method, which is used for analysis, is not dependent on specific distribution assumptions about the detection probability of birds at various perpendicular distances from the transect line. In this method density is computed from ungrouped, perpendicular distances from transect.

The four major assumptions in the line transect method are as following.

1. Birds on the line or very near to it will always be detected.
2. The birds are not moving in response to the observer and no bird is counted twice.
3. All perpendicular distances are recorded without measurement error.
4. Sighting of each bird is a statistically independent event.

The above assumptions were followed while censusing birds. The density of birds was computed by the software "DISTANCE".

The steps in the density (D) estimation by Fourier series are described in Burnham *et al.* (1980). Ungrouped data was used for analysis. Flocks were considered as single and only one perpendicular distance to the middle of the flock was measured. Density of flocks was estimated and actual density was arrived by multiplying with mean flock size. A birdcall was considered to be equivalent to a single individual and was used along with sighting record for the density estimation. The density estimation was done at two levels.

1. Total bird density was calculated for each month pooling data of all species.
2. Total bird density for two seasons was calculated pooling data of all species.

Sighting distance (Kelker's Belt)

The frequency of different sighting distance classes (5 m classes) of birds were plotted to measure the optimum sighting distance. This analysis was done using the first year's data. Forty six percentage of the birds were observed within five meters distance from the transect in the ecotone area (Fig. 3.1). Six per cent of the birds accounted for the distance from 5 - 25 m. As the distance from the transect increased, detectability of birds also increased and no birds were seen or heard at a distance more than 40 m due to the increased density of the vegetation. At Mukkali also similar pattern of bird distribution was observed. The first 20 m distance accounted for 91 per cent of birds observed. After this there was a sudden drop in detectability. This analysis showed that birds present within a distance of 40 m were only visible from transect.

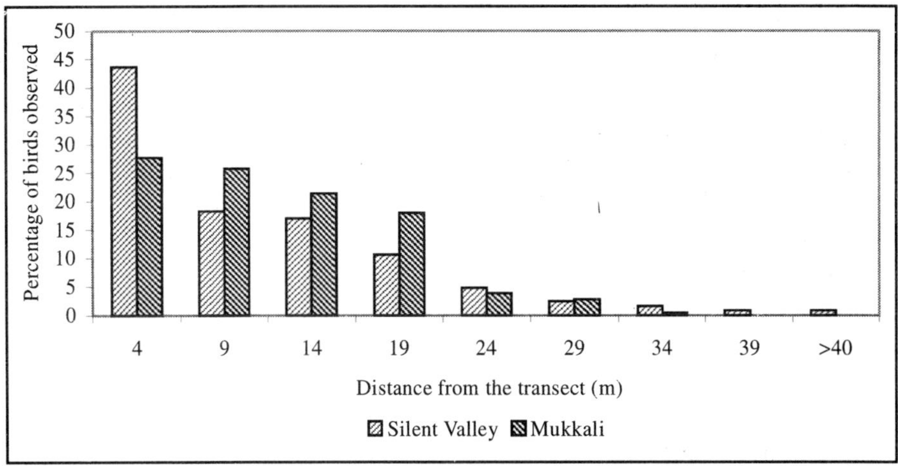

Fig. 3.1: Distribution of sighting distance

Species richness indices

Species richness indices like Margalef index and Menhinick index were calculated for both sites using the following formula (Magurran, 1988). These indices take mainly into consideration, S (Number of species recorded) and N (Total number of individuals summed over all S species).

The formula for Margalef diversity index is

$$R1 = \frac{S - 1}{Ln(n)}$$

Formula for Menhinick's index is

$$R2 = \frac{S}{\sqrt{n}}$$

Distribution models

Species of birds were ranked in their order of abundance as represented by individuals seen in each species and this was plotted in the decreasing order, for all species against the number of individuals for the two areas. The data were described using a truncated lognormal distribution (Pielou, 1975).

Diversity indices

Several diversity measures are available for assessment of species diversity in an area. Shannon index, Simpson index and Hill's diversity numbers N1 and N2 were calculated for Silent Valley and Mukkali. Diversity indices were calculated using the programme SPDIVERS.BAS (Ludwig and Reynolds, 1988).

Shannon-Weiner Index (H')

Shannon-Weiner Index of general diversity (H') is given as

$$H' = -\Sigma P_i \log P_i$$

where Pi the proportional abundance of the ith species = (n_i/N)

Simpson's Index (λ)

The following equation is used to calculate the Simpson's Index (λ)

$$\lambda = S (n_i (n_i-1)/(N (N-1))$$

where

n_i = the number of individuals in the ith species

N = total number of individuals

Hill's diversity

Hill's diversity N1 is calculated from Shannon-Weiner Index

$$N1 = e^{H'}$$

In addition, Hill's diversity N2 is calculated from Simpson's Index

$$N2 = 1/ \ddot{e}$$

Evenness indices

Five evenness measures were calculated namely Shannon evenness, Sheldon evenness and Heip evenness, Hill's evenness (E4) and Hill's evenness (E5). All the above indices were calculated using the computer programme "SPDIVERS.BAS" (Ludwig and Reynolds, 1988). The following formulae were used for calculating five Evenness measures based on Shannon index and Simpson index.

$$\text{Shannon Evenness (E1)} = \frac{H'}{\log (S)}$$

where H' = Shannon-Weiner Index

S = Number of species

$$\text{Sheldon Evenness (E2)} = \frac{e^{H'}}{S}$$

Similarity measures

Three similarity indices between the two areas were calculated namely Jaccard index, Sorenson index and modified Sorenson quantitative (Magurran, 1988).

Jaccard Index

$$C_j = j / (a + b - j)$$

where

j = the number of species common to both sites
a = the number of species in site A and
b = the number of species in site B

Sorenson Index

$$C_S = 2j / (a + b)$$

where

j = the number of species common to both sites
a = the number of species in site A and
b = the number of species in site B

Abundance

The mean number of birds seen in each month over the years is presented in the Table 3.1. The number of birds ranged from 43 to 153 at Silent Valley and from 43 to 78 at Mukkali. A slight reduction in the total number of birds was seen during the months of monsoon.

Table 3.1: Mean number of birds recorded in each month (1988-1993)

Area	Months											
	J	F	M	A	M	J	J	A	S	O	N	D
Silent Valley	91	87	70	65	76	43	46	47	95	81	109	153
Mukkali	66	54	77	52	49	66	43	51	68	58	78	41

Chi-square test was done for both places to find out any significant difference existed in total number of birds in various months. Results showed significant difference for the Silent Valley (χ^2 = 131.09; P = 0.001; df = 11) and Mukkali (χ^2 = 28.69; P = 0.01; df = 11).

Density

Mean density of birds in each month over the years is presented in Fig. 3.2. Maximum density was found in the month of December and minimum in August at Silent Valley. At Mukkali minimum density was found in the month of August and maximum in September. Over all density of birds at Silent Valley was 1,122 birds/km^2 and at Mukkali 780 birds/km^2.

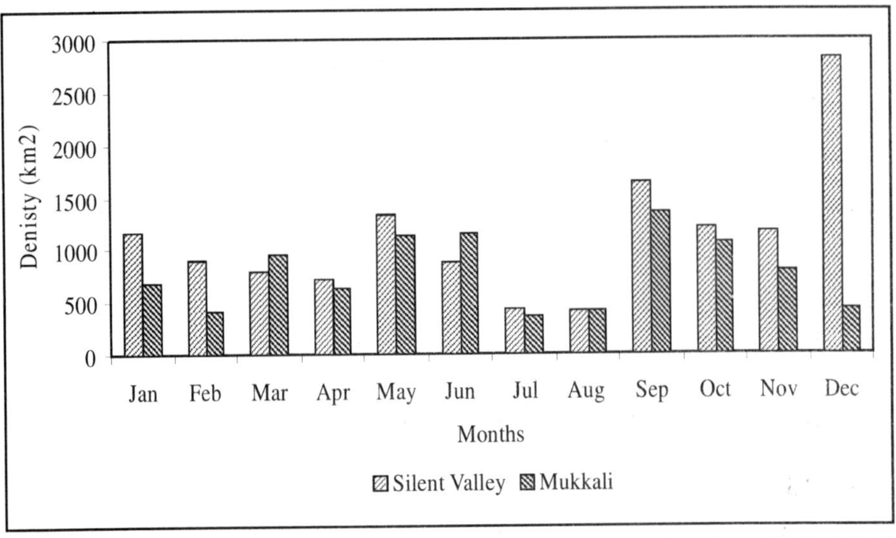

Fig. 3.2: Density of birds in each month at Silent Valley and Mukkali (1988-1993)

Individual species abundance

The abundance of common species at each site is given in Tables 3.2 and 3.3. Out of the 99 species observed at Silent Valley, monthly abundance of 10 species was calculated. Among the 96 species at Mukkali monthly abundance of 10 species was calculated.

The birds of Silent Valley can be grouped into two, based on the difference in abundance over months. The first group of resident birds showed almost a stable abundance while the second group registered an increase in abundance during summer. The first group comprises the Grey Jungle Fowl, Malabar Whistling Thrush, Southern Tree Pie, Yellow browed Bulbul, Small Green Barbet and Pied Bush Chat.

Table 3.2: Mean monthly abundance of common birds at Silent Valley

S. No.	Species name	Months											
		J	F	M	A	M	J	J	A	S	O	N	D
1.	Black Bulbul	3	5	3	4	1	-	-	-	4	-	23	36
2.	Pied Bush Chat	3	3	4	3	3	2	2	2	2	1	2	6
3.	Golden-backed Woodpecker	1	1	2	1	1	0	1	1	1	2	3	1
4.	Grey Jungle Fowl	3	2	2	1	-	-	-	1	2	2	1	1
5.	Hill Myna	6	8	5	6	5	1	-	-	14	1	7	8
6.	Malabar Whistling Thrush	1	1	1	2	3	1	2	1	0	2	1	1
7.	Red-whiskered Bulbul	1	3	4	2	4	0	1	0	3	7	1	1
8.	Small Green Barbet	2	5	4	2	1	0	0	1	2	4	6	6
9.	Southern Tree Pie	1	0	1	1	3	1	1	0	0	2	1	2
10.	Yellow-browed Bulbul	13	13	9	12	16	19	16	14	18	16	12	12

- = not recorded

Table 3.3: Mean monthly abundance of common birds at Mukkali

S. No.	Species name	Months											
		J	F	M	A	M	J	J	A	S	O	N	D
1.	Black Drongo	2	3	5	1	3	4	4	-	3	3	3	4
2.	Blossom-headed Parakeet	4	3	1	1	-	-	-	-	1	6	6	3
3.	Golden-backed Woodpecker	2	2	1	1	1	2	1	-	1	1	1	-
4.	Jungle Babbler	7	7	11	11	8	13	10	12	13	10	1	4
5.	Magpie-robin	1	2	1	2	1	1	-	-	2	-	-	-
6.	Racket-tailed Drongo	1	1	2	1	2	4	1	2	2	1	2	2
7.	Red-vented Bulbul	2	1	4	3	2	2	3	1	3	1	3	2
8.	Small Green Barbet	4	4	6	5	1	2	1	-	1	1	3	1
9.	Spotted Dove	3	2	2	1	-	-	1	-	1	2	1	-
10.	Yellow-browed Bulbul	2	1	2	1	-	2	3	3	1	2	3	2

- = not recorded

The second group is of Black Bulbul, Parakeets, Doves and Pigeons, which showed an increase during summer and a decrease during monsoon while in some months they were absent. The Small Green Barbet, Roseringed and Blossom-headed Parakeets showed maximum density during the dry months in the second area. Compared to the Silent Valley, the overall abundance of common birds was lower at Mukkali and higher during winter. Certain species showed consistent abundance in both areas. The Yellowbrowed Bulbul was stable in abundance at Silent Valley, while Red-vented Bulbul and Small Green Barbet showed a stable abundance at Mukkali.

Species richness indices

Another way of representing species richness is through the species richness indices. Margalef index and Menhinick index were calculated for both the areas (Table 3.4). These indices provide easily understandable measures of diversity.

Table 3.4: Species richness indices at Silent Valley and Mukkali

Indices	Silent Valley	Mukkali
Margalef index	11.40	12.18
Menhinick index	1.35	1.89

Margalef index and Menhinick index showed higher values for Mukkali. This suggests that the bird community at Mukkali is more diverse. Similar values were obtained for both indices from Silent Valley.

Rarefaction

Since the sample sizes from the two areas are not equal the rarefaction using Hurlbert's (1971) formula.

$$E(S) = \sum \left\{ 1 - \left[\binom{N - n_i}{n} \Big/ \binom{N}{n} \right] \right\}$$

where

E(s) = the expected number of species in the rarefied sample

n = standardized sample size

N = the total number of individuals recorded in the sample to be rarefied.

N1 = the number of individuals in the ith species in the sample.

The programme RAREFRAC.BAS was used for computation (Ludwig and Reynolds, 1988). The standardised sample size (n) is taken as the sample size of Mukkali (2628), which is the smallest. For this sample size expected number of species at Silent Valley would be 83. This indicates that Mukkali is more diverse than Silent Valley.

Distribution Models

Another way of describing diversity in a community is through species-abundance or distribution models (Fischer *et al.*, 1943). A species-abundance model utilises all the information gathered in a community and is the most complete mathematical description of the data (Magurran, 1988). Analyses were also carried out to see whether the truncated lognormal model suited the bird community at the study area. This distribution indicated the absence of a single dominant species or group of species and the presence of a long series of very rare species at Silent Valley and Mukkali. The species is represented, by less than 2 per cent of individuals recorded can be termed as rare. The observed and expected number of species was compared using χ^2 goodness of fit test. The test showed that there is no significant difference between the observed and expected distribution. This indicated that the distribution pattern of species is following truncated lognormal (χ^2 = 8.63; P = 0.30) at Silent Valley (Table 3.5). At Mukkali also the distribution pattern was following a truncated lognormal distribution (χ^2 = 9.67; P = 0.16) (Table 3.6).

Diversity indices

Species diversity

Even though species-abundance model provides full description of diversity, they cannot be compared by means of diversity indices. Indices based on the proportional abundance of species provide another approach to measure diversity.

Table 3.5: Truncated lognormal distribution at Silent Valley (χ^2 test)

Class	Upper boundary	Observed species	Expected species	χ^2
Behind veil line	0.5	-	16.13	-
1	2.5	30	24.82	1.08
2	4.5	10	11.57	0.213
3	8.5	9	12.82	1.14
4	16.5	12	12.63	0.03
5	32.5	9	11.42	0.51
6	64.5	8	9.09	0.13
7	128.5	11	6.65	2.85
8	256.5	7	4.53	1.35
9	512.5	2	2.66	0.16
10		1	2.81	1.17
Total	**99**	**115.13**	**8.63**	

Table 3.6: Truncated lognormal distribution at Mukkali (χ^2 test)

Class	Upper boundary	Observed species	Expected species	χ^2
Behind veil line	0.5	-	12.91	-
1	2.5	29	28.66	0.004
2	4.5	6	13.72	4.34
3	8.5	15	15.88	0.05
4	16.5	13	14.94	0.25
5	32.5	10	12.10	0.36
6	64.5	12	8.86	1.11
7	128.5	7	12.40	2.35
8	256.5	3	3.05	0.36
9		1	2.44	0.85
Total		**96**	**124.96**	**9.67**

They are called heterogeneity indices because they take both evenness and species richness into account (Peet, 1974). Most widely used measures of diversity indices are based on the assumption that diversity or information in a natural system can be measured the same way as that of information contained in a code or message. Shannon index of diversity, Simpson index of diversity and Hill's numbers N1 and N2 are used here.

Hill's diversity numbers which are in units of number of species, measure the "effective number of species" present in a sample. This index is calculated because N1 and N2 are suitable for addressing "any question that a heterogeneity index can answer" (Peet, 1974). Other diversity indices are merely variants of N1 and N2. Values of four diversity indices obtained for Silent Valley and Mukkali are given in Table 3.7.

Table 3.7: Bird community diversity in the study area

	No. of species	No. of individuals	Simpson's Index	Shannon index	Hill's Number N1	Hill's Number N2
Silent Valley	99	5412	0.07	3.30	27.14	14.54
Mukkali	96	2641	0.06	3.45	31.38	15.67

In order to find out whether any significant difference existed in the bird diversity between the two places a 't' test was done using the method described by Magurran (1988). For this Shannon index was used. The variance of the diversity in the two areas was calculated.

Var H′ 1 (Silent Valley) = 0.0003180924

Var H′ 2 (Mukkali) = 0.0006693683

Following formula is used to do the 't' test:

$$t = \frac{H'1 - H'2}{(\text{Var } H'1 + \text{Var } H'2)\ 1/2}$$

$$t = -4.7734$$

Significant difference was obtained in the diversity of the two places (t = - 4.7734; P = 0.05; df = 6094). This suggested that Mukkali also supported bird diversity similar to evergreen forests. Jack-knifing of diversity index was not attempted since the two diversities showed significant difference. Mukkali is a moist deciduous forest in a state of disclimax. It is likely that the colonisation by man has diversified the food resources available to birds.

Evenness indices

A variety of evenness measures are available. Two of them are based on Shannon diversity and Simpson diversity. This index measures the evenness of species-abundance and is complimentary to the diversity index concept and is a measure of how the individuals are appropriated among the species. The ratio of observed diversity to maximum diversity is taken as measure of Evenness (E). Five different evenness measures were calculated. The values of different evenness measures computed for Silent Valley and Mukkali is given in the Table 3.8.

Table 3.8: Values of different evenness measures at Silent Valley and Mukkali

Evenness Indices	Silent Valley	Mukkali
Shannon Evenness (E1)	0.72	0.75
Sheldon Evenness (E2)	0.27	0.32
Heip Evenness (E3)	0.27	0.31
Hill's Evenness (E4)	0.54	0.49
Hill's Evenness (E5)	0.52	0.48

Similarity measures

Similarity indices, another measure of diversity, were also calculated using Jaccard index and Sorenson index. Both the above-mentioned indices use qualitative data.

j = 56; a = 99; b = 96

Jaccard C_j = 0.40

Sorenson C_s = 0.57

When the similarity is complete the indices will show the value 1 and 0 in the case of site being dissimilar with no common species. The computed similarity indices show medium similarity between the two study sites. Out of 137 species recorded in the two places only 56 species were common to both sites.

One disadvantage with the above indices is that they are not considering abundance data. Instead the presence of individuals is considered equally in the equation whether they are abundant or rare. Similarity measures based on quantitative data solve this problem. Modified Sorenson quantitative is such a measure.

Sorenson quantitative C_n = 2jn/(aN+bN)

aN = the total number of individuals in site A;

bN = the total number of individuals in site B;

jn = the sum of the lower of the two abundances recorded for species found in both sites.

aN = 5412

bN = 2641

jn = 1750

Sorenson quantitative C_n = 0.44

Of the three similarity indices computed, Jaccard and Sorenson quantitative shows a similarity above 40 per cent and the Sorenson index shows a similarity of 57 per cent between Silent Valley and Mukkali.

Comparison with other tropical forests

In the present study 1,122 individuals/km^2 were recorded at Silent Valley and 780 individuals/km^2 at Mukkali. The above values are comparable to a number of similar studies on the composition and abundance of tropical forest birds. Karr (1971) estimated a population density of 1820 pairs of birds/km^2 in Panama's tropical forests. Thiollay (1986) reported a density of 760 pairs/km^2 for a total of 263 species in French Guiana. A long-term study conducted by Brosset (1990) in Gabon on 364 species of birds produced a density of 3,690 individuals/km^2. A study by Bell (1983) on birds of lowland rainforest in New Guinea produced a density of 3450 pairs/km^2. A study by Terborgh *et al.* (1990) on 245 species of Amazonian forest bird community had shown a density of 1910 individuals/km^2.

Abundance and density

The main drawback of the census method employed is the under estimation of density either because the bird moves away from the observer before being located or because birds are overlooked. It is reported that by using fixed width line transect, there is a chance to miss at least 50 per cent of density (Franzeb, 1981 and Hilden, 1981). Bell and Ferrier (1985) reported that transect census underestimated densities of many species but that those of variable-width (Emlen, 1971) are reliable, than fixed-width transects.

A drop in bird abundance during monsoon season with adverse climate as observed in this study was reported by Ramakrishnan (1983) and Jayson (1989). Similarly Morison *et al.* (1980) reported reduction of birds during non-winter period and their increase during winter. One factor, which may influence the abundance, is detectability. Seasonal differences in detectability are common for most of the bird species (Emlen, 1971). These differences result from changes in weather and habitat structure. Increasing foliage density decreased the visibility of birds. But in the study area foliage structure was identical in all seasons and only rainfall had some influence on detectability.

The higher density (1,122 birds/km^2) observed at Silent Valley may be due to the vegetation with more canopy layers. Frugivorous species like Black Bulbul, Parakeets and Doves were not sighted during major part of the wet season. This local migration of some bird species from the area may be to avoid harsh climate coupled with lack of food materials (Jayson, 1989).

Species richness indices

Species richness in an area is dependent on the availability of food resource, climate of the area, evolutionary history, and predation pressure. Species richness indices, diversity indices and rarefaction showed high diversity for Mukkali. Tree diversity index was higher at Silent Valley (2.91) and low at Mukkali (2.57). But bird diversity indices showed significant difference. This indicated the non-existence of any relationship between tree diversity and bird diversity.

Species-abundance relationships

A large number of rare species of birds occur at Silent Valley and Mukkali. This is typical of tropical forests (Lovejoy, 1975). The numbers of factors, which control the species richness in an area, are broadly divided into the historical and the ecological (Giller, 1984). Among the historical factors, speciation and the crossing of geological barriers and supply of colonists are more important. Among the ecological factors mortality due to predation is an important factor. Many cases of mortality due to predation were recorded from the area.

Currently four models are available for describing species-abundance distribution. They are the geometric series, the lognormal, the log series and Mac Arthur's Broken-stick Model. Preston (1948) introduced the lognormal distribution to explain the species-abundance data. Usually in ecological work distribution of species is always truncated at the left side (Preston, 1962). Geometric series patterns are usually found in species-poor or harsh environments. Log-series patterns are usually observed where one or a few factors dominate the ecology of a community. Lognormal distribution is found in most of the biological populations. The Broken-stick model distribution shows the maximum equitable distribution of available resources.

Species-abundance distribution at Silent Valley and Mukkali follows truncated lognormal model (Jayson and Mathew, 1993). This study also showed that the bird community in the study area is following truncated lognormal distribution. It is reported that the Amazonian forest bird community also showed lognormal distribution in species-abundance (Terborgh *et al.*, 1990). As in the case of birds species-abundance of ants in Kobbeduinen and Kooiduinen approximately agreed with lognormal distribution (Boomsa and Van Loon, 1982).

Diversity indices

Diversity means either the total number of species in a community or the dual concept of diversity combining both species richness and relative abundance (Peet, 1974). According to Odum (1974) the ratio between the number of species and "importance values" (Numbers) of individuals is called species diversity. Diversity indices are basically dependent on two factors, one is species richness and another one is evenness. Considerable discussion is going on around the measurement of diversity. It is directly correlated with the stability of the ecosystems and will be higher in the biologically controlled systems and will be low in polluted ecosystems (Rosenberg, 1976). But some authors like Hurlbert (1971) even consider diversity indices as a 'non-concept'.

In the Silent Valley and Mukkali diversity indices showed closely similar values. There was slightly higher evenness at Mukkali. This shows that Silent Valley has more rare avian species than Mukkali. This is quite natural, as tropical wet evergreen forests are known to support more rare species than other habitats. Similar observations have been also reported in other studies (Pearson, 1977). As the evenness measures show high values, it can be concluded that species are uniformly represented by individuals. Similarly, measures based on Simpson diversity and Hill's evenness E4 and E5 also show the same trend. These measures are the ratio of numbers of very abundant to abundant species. E5 is known as "Modified Hill's ratio". E5 approaches zero as a single species becomes more and more dominant in a community (Alatalo, 1981). Evenness measure (E5) also indicates that the species at Mukkali are more uniformly represented than at Silent Valley.

A number of hypotheses have been forwarded to explain the reasons for the characteristic diversity profiles of different habitats. Habitat heterogneity, in addition to area, is an important determinant of species richness (Boecklen and Simberloff, 1986). Habitat factors such as tree density, basal area, number of tree species, percent ground cover, per cent canopy cover and canopy height, are also important in determining diversity. At the Silent Valley bird diversity was possibly related to the history of the area, foliage density and local migration of the species. Habitat heterogeneity at Mukkali may be one of the factors for the higher diversity recorded from that area.

Diversity indices are extensively used in environmental monitoring and testing and in conservation. As the objective of the world conservation strategy is to maximise diversity of habitats, these indices are extensively used to monitor and evaluate habitats. According to a study conducted by Usher (1986) among the criteria used for evaluation of conservation schemes, diversity is the most frequently adopted criteria. Diversity indices are directly correlated with the stability of the ecosystem and will be high in biologically controlled systems. Diversity indices at Silent Valley showed the same. All diversity indices have limitations because they attempt to combine a number of variables that characterise community structure. These measures are easy to compute, but difficult to interpret.

❑❑❑

Chapter 4

Seasonal Changes of the Community

Tropical forests support a stable population of birds throughout the seasons in comparison to the temperate forests, where there is marked variation in tune with change of seasons (Wright, 1970). The seasonal changes of bird community at Silent Valley and Mukkali are presented in this chapter. There are two seasons namely monsoon (June to November) and summer (December to May). The availability of food at Silent Valley was studied to find out the relationship between seasonal changes of the community and the abundance of food.

METHODS

Seasonal changes

Occurrence of birds in each month obtained from the census was used for seasonality analysis. Seasonal index of birds for each month was calculated using Time Series Analysis by the method of Simple Averages (Rao, 1983). Mean data obtained from the study period was used for this analysis.

Factors governing changes in population

Bird community parameters were correlated with climatic factors like rainfall and temperature.

Foraging guilds

Food items of all recorded bird species of Silent Valley and Mukkali were gathered from the literature (Ali and Ripley, 1983) and from personal observations. Foraging guild composition in each season was compiled from these data.

Availability of food

In addition to this, phenological observations were compared with bird community parameters. To quantify the availability of insect food, insects on the ground, among herbs and grass were collected using sweep net every month. Every month one and half hours were spent for sweep netting. This was done in the same direction and in the same place in each month. All the insects collected were identified up to Orders.

Climate

Heavy rainfall was recorded at Silent Valley during the monsoon season ranging from a total of 803 mm to 2043 mm within a month (Table 4.1). December, January, February and March practically had no rain. Temperature also comes down during the months of monsoon (Table 4.2). Compared to Silent Valley the quantum of rainfall was less at Mukkali during the monsoon season (628 mm to 1623 mm, Table 4.3). Temperature at Mukkali also decreased during monsoon (Table 4.4).

Species composition

Most common bird species, which were recorded during the months of monsoon from Silent Valley and Mukkali are given in the Table 4.5.

Table 4.1: Rainfall data of Silent Valley

Months	Monthly Rainfall (mm)					
	1988	1989	1990	1991	1992	1993
January	-	-	-	-	1.6	-
February	-	-	-	-	-	20.2
March	-	-	6.2	11.4	-	14.4
April	-	72.6	13.4	117.7	120.8	6.2
May	-	70.2	803.1	30.8	249.7	-
June	-	888.7	836.4	1421.6	1532.7	-
July	1045.2	1642.0	954.8	1888.4	2043.8	-
August	1317.9	714.0	1069.2	593.4	1102.8	-
September	741.3	841.0	344.6	143.8	692.8	-
October	206.3	534.0	256.2	422.3	379.4	-
November	48.5	64.8	86.7	127.2	307.3	-
December	-	2.0	-	-	-	-

Table 4.2: Temperature and humidity at Silent Valley

Months	Mean temperature °C	Mean RH (%)
January	22	88
February	21	86
March	19	76
April	20	86
May	20	84
June	19	89
July	19	89
August	20	90
September	21	89
October	20	88
November	22	88
December	22	88

Table 4.3: Rainfall data of Mukkali

Months	Monthly Rainfall (mm)					
	1988	1989	1990	1991	1992	1993
January	-	-	29.0	20.6	-	-
February	-	-	-	2.6	-	6.0
March	44.0	40.0	45.4	4.4	-	3.6
April	116.0	93.9	57.8	130.2	41.2	-
May	263.0	71.0	628.6	42.8	110.4	185.5
June	1055.0	928.0	579.6	1412.2	895.8	-
July	1107.0	1259.2	648.2	1623.4	1321.0	-
August	984.0	229.1	672.0	406.0	666.6	-
September	902.0	531.9	88.4	224.8	234.8	-
October	317.0	521.8	455.6	712.1	682.6	-
November	43.0	146.8	90.8	431.0	170.4	-
December	-	27.4	29.0	-	-	-

Table 4.4: Temperature recorded at Mukkali ºC

Months	Year						Mean
	1988	1989	1990	1991	1992	1993	
January	-	21	21	22	24	25	23
February	-	21	23	23	26	26	24
March	-	23	24	26	22	27	24
April	-	27	27	27	27	27	27
May	-	26	25	30	26	26	27
June	24	25	23	23	27	-	24
July	23	24	23	22	24	-	23
August	23	22	22	23	23	-	23
September	24	23	24	25	25	-	24
October	22	23	25	25	25	-	24
November	23	23	23	25	25	-	24
December	22	22	22	23	26	-	23

Table 4.5: Common species of birds recorded from Silent Valley and Mukkali during monsoon.

Species	Silent Valley	Mukkali
Painted Bush Quail	√	
Grey Jungle Fowl	√	
Spotted Dove		√
Blossom-headed Parakeet	√	√
White-breasted Kingfisher		√
Small Green Barbet	√	√
Golden-backed Woodpecker	√	√
Black Drongo		√
Bronzed Drongo		√
Racket-tailed Swallow	√	√
Hill Myna	√	
Tree Pie		√
Southern Tree Pie	√	√
House Crow		√
Scarlet Minivet	√	√
Red-whiskered Bulbul	√	√
Red-vented Bulbul		√
Yellow-browed Bulbul	√	√
Jungle Babbler	√	√
Magpie Robin		√
Pied Bush Chat	√	
Malabar Whistling Thrush	√	√
Nilgiri White-eye	√	

Species of birds recorded during summer months at Silent Valley and Mukkali are given in Table 4.6.

Table 4.6: Common species of birds recorded from Silent Valley and Mukkali during summer months

Species	Silent Valley	Mukkali
Blackwinged Kite	√	√
Painted Bush Quail	√	
Grey Junglefowl	√	√
Spotted Dove		√
Greyfronted Green Pigeon	√	
Yellowlegged Green Pigeon	√	
Imperial Pigeon	√	
Emerald Dove	√	
Roseringed Parakeet	√	√
Blossomheaded Parakeet	√	√
Bluewinged Parakeet	√	√
Malabar Lorikeet		√
Malabar Grey Hornbill	√	√
Small Green Barbet	√	√
Goldenbacked Woodpecker	√	√
Redrumped Swallow	√	
Blackheaded Oriole	√	√
Black Drongo	√	√
Racket-tailed Drongo	√	√
Common Myna		√
Hill Myna	√	
Tree Pie		√
Southern Tree Pie	√	√
House Crow		√
Scarlet Minivet	√	√
Redwhiskered Bulbul	√	√
Redvented Bulbul		√
Yellowbrowed Bulbul	√	√
Black Bulbul	√	√
Jungle Babbler	√	√

Contd...

Tickell's Blue Flycatcher	√	
Leaf Warbler	√	
Pied Bush Chat	√	√
Magpie-robin		√
Malabar Whistling Thrush	√	√
Whitethroated Ground Thrush		√
Yellow Wagtail	√	√
Tickell's Flowerpecker	√	
Purplerumped Sunbird		√
Small Sunbird	√	√
Common Rose Finch	√	

Seasonal changes

The total number of birds, monthly density and species richness of birds at Silent Valley and Mukkali underwent slight reduction during monsoon season and increase during the summer months. It appeared that the main reason for the drop in number of birds and species during monsoon months is the local movements of species to avoid harsh climate. There is suggestive evidence in the fact that there is a fall in species richness during the months of monsoon, which indirectly contributes to the decrease in total bird abundance. Species like Black Bulbul, Emerald Dove and Imperial Pigeon were practically absent during this season at Silent Valley. Local movements in search of optimum habitats are possible due to the availability of other habitats in the vicinity. The tracts where this study was conducted were only fragmented forest patches and separated from other forest types by a distance of about 20 km. During the monsoon moths Silent Valley had winds of high velocity, which caused uprooting of many trees. This also could have disturbed the birds considerably.

Seasonal index

The percentage of increase or decrease of birds in each month is obtained from this analysis. An index of 80 for March means that the variable under study is typically 20 per cent below the annual average (base = 100) and an index of 106 for June means that the variable is typically 6 per cent above the annual average. The results showed that in six months namely September, October, November, December, January and February, more birds were present than the annual average at Silent Valley. In all other months birds were below annual average (Table 4.7). In the case of Mukkali more birds than the annual average was found in the months of September, November, January and March. Maximum seasonal index was obtained in the month of November.

Table 4.7: Seasonal index of birds present in each month at Silent Valley and Mukkali

	Months											
	J	F	M	A	M	J	J	A	S	O	N	D
Silent Valley	114	109	88	81	95	54	58	59	119	101	136	153
Mukkali	113	92	131	89	84	113	73	87	116	99	133	70

Seasonal change in abundance

Data were pooled into two seasons namely monsoon and summer to find out seasonal difference in the total number of birds. Chi-square test revealed significant difference in the number of birds between the seasons at Silent Valley (Table 4.8). Highest number of birds per month was observed in the 1991 summer season (91) and the lowest number was in monsoon 1992 (53). At Mukkali there was no significant difference among seasons in the total number of birds (Table 4.9).

Significant differences in the number of birds per month between Silent Valley and Mukkali were observed during three summer seasons (Table 4.9). During these seasons higher number of birds was seen at Silent Valley. But during 1992 summer there was no significant difference in the number of birds was observed between Silent Valley and Mukkali. During monsoon season also no significant difference was obtained between Silent Valley and Mukkali.

Significant difference in density was obtained between seasons in different years at Silent Valley and when we compare summer season with monsoon, a significant difference is also obtained (χ^2 = 62.25; P = 0.05; df = 1). High density was obtained during summer season at Silent Valley and Mukkali. Significant difference in density was obtained between the seasons in different years at Mukkali. Summer and monsoon seasons also showed significant difference (χ^2 = 39.33; P = 0.05, df = 1). Except for the two summer seasons significantly higher bird density was obtained at Silent Valley.

Seasonal difference in diversity

Variations in diversity of birds in different seasons at the Silent Valley and Mukkali are computed and given in Table 4.10. Shannon diversity index was computed for this purpose.

Table 4.8: Mean number of birds recorded per month in different seasons at Silent Valley and Mukkali

Seasons	Silent Valley	Mukkali	Total	χ^2	P <
Monsoon 1988	70	76	146	0.25	NS
Summer 1989	95	52	147	12.58	0.02
Monsoon 1990	74	48	122	5.50	0.02
Summer 1991	91	50	141	11.92	0.001
Monsoon 1992	53	67	120	1.63	NS
Summer 1992	83	70	153	1.10	NS
Summer 1993	89	59	148	6.08	0.02
Total	555	422			
χ^2	16.36	11.83			
P <=	0.02	NS			

Table 4.9: Seasonal variation in total bird density at Silent Valley and Mukkali

Seasons	Density/km² Silent Valley	Mukkali	Mean density Total	Mean	χ^2	P <
Monsoon 1988	1036	638	1674	837	94.63	0.001
Summer 1989	2123	1662	3785	1892.5	56.15	0.001
Monsoon 1990	685	401	1086	543	74.27	0.001
Summer 1991	741.4	370	1111.4	555.7	124.11	0.001
Monsoon 1992	493	792	1285	642.5	69.57	0.001
Summer 1992	823	757	1580	790	2.76	NS
Summer 1993	608	611	1219	609.5	.007	NS
Total	6509.40	5231				
χ^2	1976.52	1514.23				
P <	0.001	0.001				

Table 4.10: Seasonal difference in Shannon index at Silent Valley and Mukkali

Seasons	Silent Valley	Mukkali
Monsoon 1988	2.77	2.50
Monsoon 1989	2.38	2.63
Summer 1989	3.20	2.96
Monsoon 1990	2.70	2.85
Summer 1990	3.01	2.95
Summer 1991	3.23	3.08
Monsoon 1992	2.74	3.13
Summer 1992	3.29	3.46
Summer 1993	2.88	3.25

Shannon index showed high values in summer seasons ($\chi^2 = 3.12$; n = 5) and lower values during monsoon ($\chi^2 = 2.65$; n = 4) seasons at Silent Valley. Similar trend was obtained in other community parameters also. Same type of variation in diversity indices was seen at Mukkali also during different seasons (monsoon $\chi^2 = 2.78$; n = 4 and summer $\chi^2 = 3.14$; n = 5).

Factors governing changes in population

Rainfall

The main factors which were found to influence the bird communities were the rainfall and wind. A direct relationship obtained between the monthly rainfall, monthly bird population and monthly bird density and total number of bird species at Silent Valley. When the rainfall increased all the three population parameters decreased, and as the rainfall reduced population of birds again started to build up at the Silent Valley (Figs. 4.1, 4.3 and 4.5). At Mukkali also rainfall had an influence on bird community, but not on same magnitude as that of Silent Valley (Figs. 4.2, 4.4 and 4.6) was recorded.

At Silent Valley significant negative correlation obtained between mean monthly total rainfall (1988-1993) and number of species in each month (r = - 0.731; P = 0.01; n = 12). Significant correlation also obtained between mean monthly rainfall and total number of birds in each month (r = - 0.66; P = 0.05; n = 12). But no significant correlation obtained between density of birds in each month and rainfall (r = - 0.45; P = 0.05; n = 12).

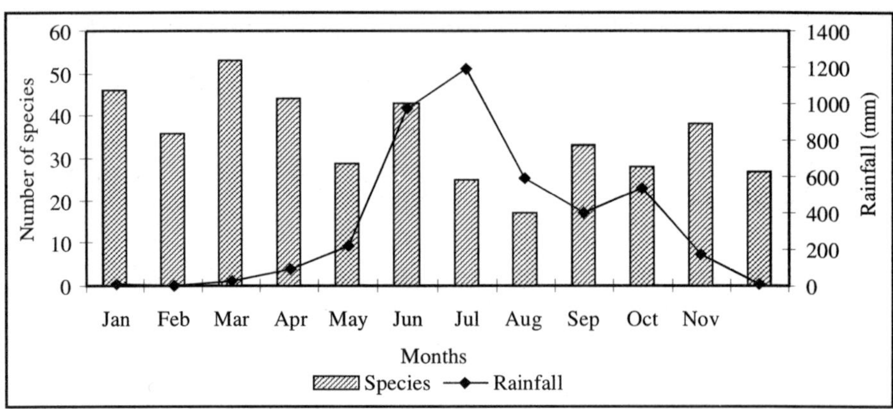

Fig. 4.1: Relation between rainfall and number of species at Silent Valley (1988-1993)

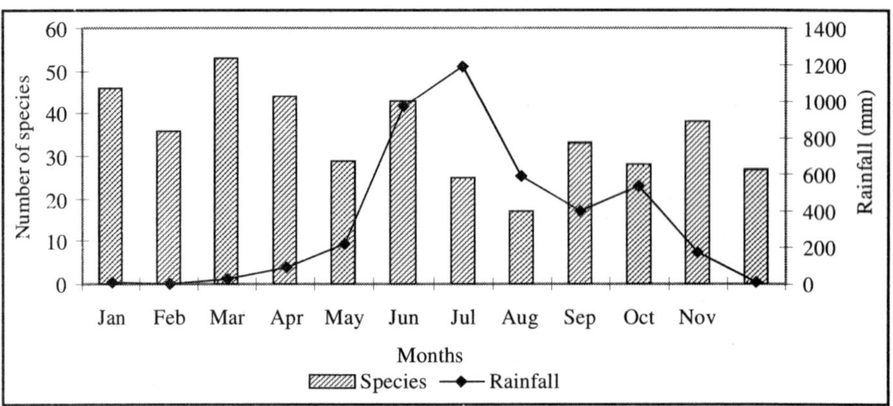

Fig. 4.2: Relation between rainfall and number of species at Mukkali (1988-1993)

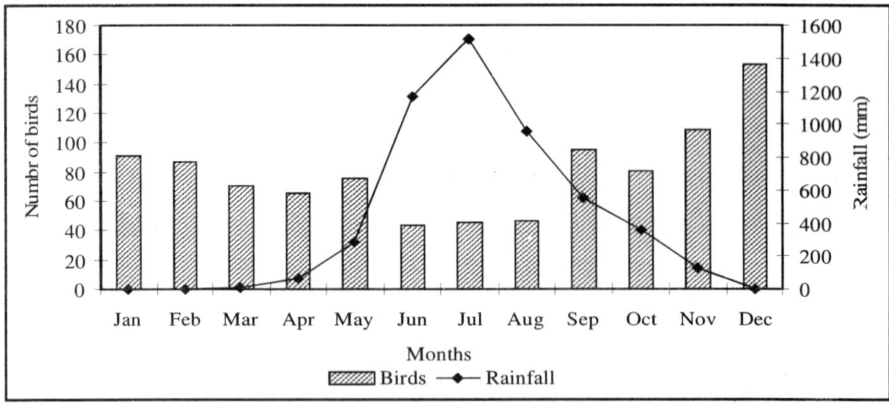

Fig. 4.3: Relation between rainfall and abundance of birds at Silent Valley (1988-1993)

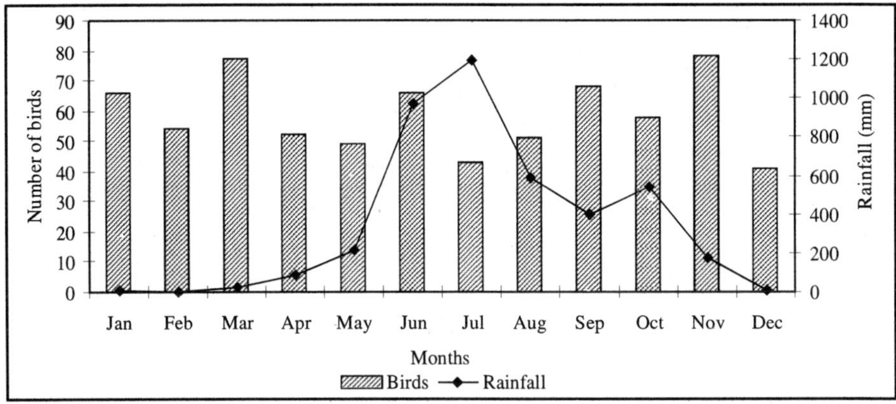

Fig. 4.4: Relation between rainfall and number of birds at Mukkali (1988-1993)

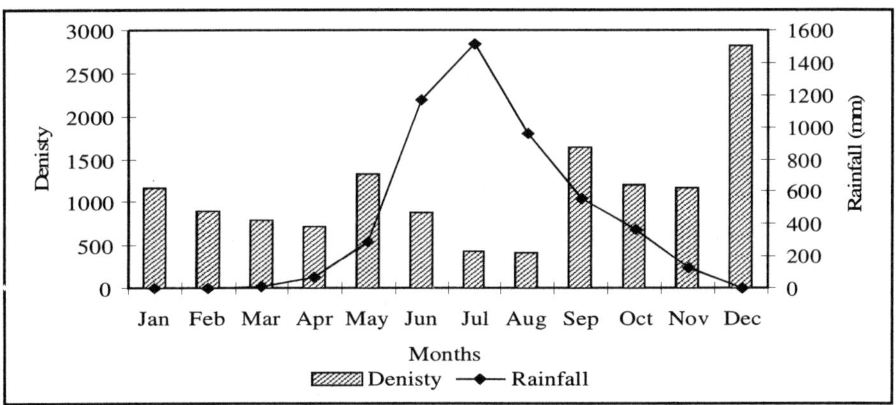

Fig. 4.5: Relation between rainfall and density of birds at Silent Valley (1988-1993)

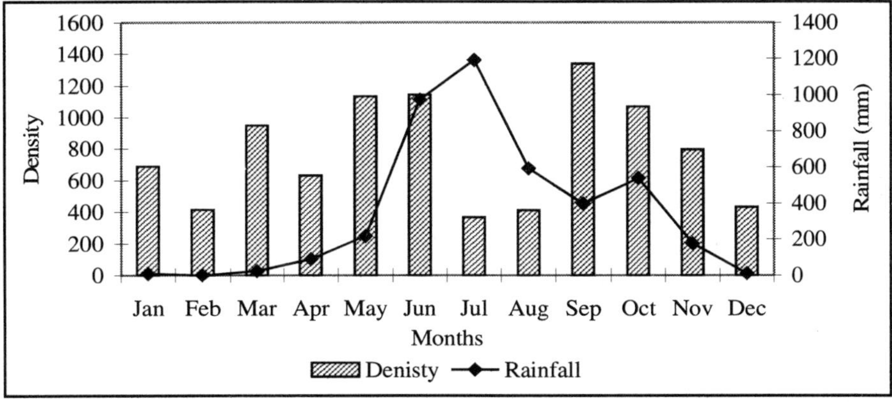

Fig. 4.6: Relation between rainfall and density of birds at Mukkali (1988-1993)

There was no significant correlation obtained between mean monthly total rainfall and community parameters from Mukkali. Here also mean monthly total rainfall was negatively correlated with the number of species (r = - 0.41; P = 0.05; n = 12). No significant correlation was obtained between mean monthly total rainfall and total number of individuals (r = - 0.21; P = 0.05; n = 12) and density of birds in each month (r = - 0.06; P = 0.05; n = 12). It shows that rainfall has no significant effect on the bird community at Mukkali.

Temperature

Significant positive correlation was obtained between the temperature and community parameters at Silent Valley. Number of species increased as the temperature raised (r = 0.57; P = 0.05; n = 12). Similarly total number of birds (r = 0.83; P = 0.05; n = 12) and density (r = 0.62; P = 0.05; n = 12) also showed an increment as the temperature became more hot. At Mukkali no such significant correlation was obtained between temperature and birds. Coefficient of correlation values were: temperature and number of species (r = 0.21; P = 0.05; n = 12); temperature and total number of birds (r = -0.08; P = 0.05; n = 12).

Seasonal changes in the foraging guilds

Silent Valley

Higher number of omnivorous species was observed during the monsoon season and insectivorous species were low during this period (Tables 4.11 to 4.20). Compared to summer season frugivorous species were also low in abundance during monsoon season (Table 4.21). But during the summer season the omnivores and insectivores do not show much difference in abundance. There is significant increase in the number of frugivorous birds between summer and monsoon seasons) $\chi^2 = 11.58$; P = 0.01). No significant difference is found in other foraging guilds. For χ^2 analysis the last seasons summer 93 was omitted to equalise the samples.

Table 4.11: Omnivorous birds recorded at Silent Valley during different seasons

Species	Mon. 88	Mon. 89	Mon. 90	Mon. 92	Sum. 89	Sum. 90	Sum. 91	Sum. 92	Sum. 93
Red Spurfowl			√				√		
Grey Junglefowl	√	√	√	√	√	√	√	√	√
Malabar Trogan								√	
Malabar Grey Hornbill	√	√	√	√		√	√	√	√
Great Indian Hornbill		√					√	√	√
Goldenbacked Woodpecker	√	√							
Blacknaped Oriole					√				
Black Drongo	√		√	√	√	√	√	√	√
Whitebellied Drongo							√		
Bronzed Drongo			√			√	√		
Common Myna								√	
Tree Pie		√							
Southern Tree Pie	√	√	√	√	√	√	√	√	√
Goldenfronted Chloropsis					√			√	
Redwhiskered Bulbul	√	√	√	√	√	√	√	√	√
Redvented Bulbul	√				√			√	
Black Bulbul	√	√	√	√	√	√	√	√	√
Black bird			√			√			
Scmitar Babbler				√		√		√	√
Whiteheaded Babbler					√			√	
Pied Bush Chat		√	√	√	√	√		√	√
Malabar Whistling Thrush	√	√	√	√	√	√	√	√	√
Orangeheaded Ground Thrush						√			
Whitethroated Ground Thrush					√	√	√		√
Grey Tit						√			
Indian Trcc Pipit						√			
Whitethroated Munia					√				

Table 4.12: Insectivorous birds recorded at Silent Valley during different seasons

Species	Mon. 88	Mon. 89	Mon. 90	Mon. 92	Sum. 89	Sum. 90	Sum. 91	Sum. 92	Sum. 93
Blackwinged Kite				√	√	√	√	√	√
Brahminy Kite	√								
Redspur Fowl								√	
Redwinged Crested Cuckoo			√						
Whiterumped Spinetail Swift	√						√		
Chestnutheaded Bee-eater			√				√	√	
Goldenbacked Woodpecker			√	√	√	√	√	√	√
Goldenbacked Threetoed Woodpecker				√	√	√			
Great Black Woodpecker				√	√	√		√	
Heartpotted Woodpecker		√	√		√				√
House Swallow		√	√		√		√		√
Redrumped Swallow	√								
Golden Oriole						√			√
Racket-tailed Drongo	√	√	√	√	√	√	√	√	√
Scarlet Minivet	√		√	√	√	√		√	√
Common Iora								√	
Rubythroated Yellow Bulbul	√								
Spotted Babbler				√				√	
Rufous Babbler				√					
Brown Flycatcher								√	
Brownbreasted Flycatcher			√						
Whitebellied Blue Flycatcher			√						
Tickell's Blue Flycatcher						√	√		
Nilgiri Flycatcher			√		√		√	√	
Paradise Flycatcher			√	√	√			√	√
Leaf Warbler	√		√	√	√		√		
Magpie Robin								√	
Yellowcheeked Tit			√		√	√		√	
Velvetfronted Nuthatch			√	√	√	√		√	
Grey Wagtail	√							√	
Purplerumped Sunbird		√							
Lottens Sunbird	√							√	
Spiderhunter								√	√

Table 4.13: Frugivorous birds recorded at Silent Valley during different seasons

Species	Mon. 88	Mon. 89	Mon. 90	Mon. 92	Sum. 89	Sum. 90	Sum. 91	Sum. 92	Sum. 93
Painted Bush Quail	√		√		√	√	√	√	√
Greyfronted Green Pigeon					√	√	√	√	
Yellowlegged Green Pigeon			√		√	√	√	√	
Imperial Pigeon			√		√	√	√	√	
Bluerock Pigeon						√			
Nilgiri Wood Pigeon						√			
Spotted Dove					√	√		√	
Emerald Dove			√		√	√	√	√	√
Roseringed Parakeet		√	√	√	√		√	√	√
Blossomheaded Parakeet	√	√		√	√		√	√	√
Bluewinged Parakeet	√		√		√	√	√		√
Malabar Lorikeet	√		√		√	√	√		
Small Green Barbet	√	√	√	√	√	√	√	√	√
Blackheaded Oriole						√			
Hill Myna	√	√	√	√	√	√	√	√	√
Fairy blue Bird						√			
Yellowbroad Bulbul	√		√	√	√	√	√	√	√
Tickell's flowerpecker						√			
Nilgiri White-eye								√	
Rufousbellied Munia				√					
Spotted Munia						√			
Blackheaded Munia	√			√	√	√	√		√
Rose Finch					√	√	√	√	

Table 4.14: Carnivorous birds recorded at Silent Valley during different seasons

Species	Mon. 88	Mon. 89	Mon. 90	Mon. 92	Sum. 89	Sum. 90	Sum. 91	Sum. 92	Sum. 93
Pond Heron					√		√		
Blackwinged Kite	√								
Shikra	√			√			√		
Black Eagle		√	√				√	√	
Crested Serpent Eagle	√		√	√	√		√		
Falcon sp.							√		
Shorteared Owl							√		

Table 4.15: Graminivorous and nectarivorous birds recorded at Silent Valley during different seasons

Species	Mon. 88	Mon. 89	Mon. 90	Mon. 92	Sum. 89	Sum. 90	Sum. 91	Sum. 92	Sum. 93
Graminivorous birds									
Yellow Wagtail	√	√	√	√	√	√	√	√	√
Nectarivorous birds									
Tickell's Flowerpecker	√				√				
Small Sunbird	√	√		√	√	√	√	√	√

Table 4.16: Omnivorous birds recorded at Mukkali during different seasons

Species	Mon. 88	Mon. 89	Mon. 90	Mon. 92	Sum. 89	Sum. 90	Sum. 91	Sum. 92	Sum. 93
Pond Heron							√		
Grey Junglefowl				√	√	√	√	√	√
Malabar Trogon		√				√			
Malabar Grey Hornbill			√	√		√	√		
Great Indian Hornbill			√				√		
Goldenbacked Woodpecker	√	√	√	√	√	√	√	√	√
Golden Oriole							√		
Blackheaded Oriole		√		√		√	√	√	√
Black Drongo						√			√
Common Myna			√	√	√	√	√	√	√
Southern Tree Pie	√	√	√	√	√	√	√	√	-
Goldenfronted Chloropsis		√		√	√	√	√	√	√
Goldmantled Chloropsis			√	√		√	√	√	
Fairy Bluebird				√					
Rubythroated Yellow Bulbul		√						√	
Redwhiskered Bulbul	√	√	√	√	√	√	√		√
Yellowbrowed Bulbul		√	√	√		√	√	√	√
Black Bulbul				√	√	√	√	√	√
Scmitar Babbler				√					
Rufous Babbler				√				√	√
Jungle Babbler	√	√	√	√		√	√	√	-
Magpie-robin		√	√	√		√	√	√	√
Pied Bush Chat		√	√	√	√	√	√	√	√
Bluerock Thrush							√	√	√
Malabar Whistling Thrush				√			√	√	√
Orangeheaded Ground Thrush									√
Whitethroated Ground Thrush	√			√		√	√	√	
Grey Tit								√	
Yellow Wagtail		√			√		√		√
Pied Wagtail				√					
Small Sunbird				√					
Loten's Sunbird				√				√	
Little Spiderhunter				√					
Nilgiri White-eye				√			√	√	

Table 4.17: Insectivorous birds recorded at Mukkali during different seasons

Species	Mon. 88	Mon. 89	Mon. 90	Mon. 92	Sum. 89	Sum. 90	Sum. 91	Sum. 92	Sum. 93
Brainfever Bird								√	
Jungle Owlet					√				√
Hoopoe						√			
Chestnutheaded Bee-eater				√		√	√	√	√
Small Yellownaped Woodpecker			√				√		
Goldenbacked Threetoed Woodpecker			√	√	√				
Heartpotted Woodpecker		√				√	√		
Indian Pitta						√		√	√
House Swallow	√	√	√	√	√	√	√	√	√
Redrumped Swallow				√					
Brown Shrike		√				√	√		√
Baybacked Shrike		√		√	√			√	
Rufousbacked Shrike							√		
Black Drongo	√	√	√	√	√		√	√	
Bronzed Drongo	√	√	√	√	√		√	√	
Racket-tailed Drongo	√	√	√	√	√	√	√	√	√
Jungle Myna			√			√		√	√
Hill Myna				√			√	√	√
Smaller Grey Cuckoo Shrike					√	√			
Scarlet Minivet	√	√	√	√	√	√	√	√	√
Common Iora				√				√	
Whiteheaded Babbler	√		√	√	√			√	
Brown Flycatcher				√	√			√	
Brownbreasted Flycatcher									√
Rufoustailed Flycatcher									√
Black-and-orange Flycatcher								√	
Tickell's Blue Flycatcher							√		
Verditor Flycatcher								√	
Paradise Flycatcher				√	√		√		
Tailor Bird				√					
Leaf Warbler	√			√		√		√	
Black Bird							√		
Yellowcheeked Tit				√		√		√	√
Velvetfronted Nuthatch		√		√					
Forest Wagtail							√	√	√
Yellow Wagtail				√				√	

Table 4.18: Frugivorous birds recorded at Mukkali during different seasons

Species	Mon. 88	Mon. 89	Mon. 90	Mon. 92	Sum. 89	Sum. 90	Sum. 91	Sum. 92	Sum. 93
Painted Bush Quail							√		
Greyfronted Green Pigeon					√		√		
Yellowlegged Green Pigeon							√		
Imperial Pigeon							√		
Emerald Dove							√		
Roseringed Parakeet			√	√	√		√	√	√
Blossomheaded Parakeet	√		√	√	√		√	√	√
Bluewinged Parakeet			√				√	√	√
Malabar Lorikeet				√	√	√	√	√	
Indian Koel						√			
Small Green Barbet	√	√	√	√	√	√	√	√	√
Redvented Bulbul	√	√	√	√	√	√	√	√	√
Rose Finch							√		

Table 4.19: Carnivorous birds recorded at Mukkali during different seasons

Species	Mon. 88	Mon. 89	Mon. 90	Mon. 92	Sum. 89	Sum. 90	Sum. 91	Sum. 92	Sum. 93
Blackwinged Kite			√	√			√	√	√
Shikra							√		
Black Eagle				√	√		√		
Crested Serpent Eagle			√	√		√	√		
Falcon sp.				√					
Crow Pheasant					√		√		
Shorteared Owl							√		
Grey Shrike								√	
Tree Pie	√		√	√		√	√	√	√

Table 4.20: Graminivorous, Nctarivorous and Piscivorous birds recorded at Mukkali during different seasons

Species	Mon. 88	Mon. 89	Mon. 90	Mon. 92	Sum. 89	Sum. 90	Sum. 91	Sum. 92	Sum. 93
Graminivorous birds									
Spotted Dove			√	√		√	√	√	√
Rufousbellied Munia				√		√	√		
Blackheaded Munia				√		√	√	√	√
Nectarivorous birds									
Haircrested Drongo			√						
Flowerpecker	√		√		√				
Purplerumped Sunbird				√				√	
Small Sunbird				√	√	√	√	√	√
Piscivorous birds									
Whitebreasted Kingfisher				√		√		√	

Table 4.21: Foraging guild composition in different seasons at Silent Valley (Percentages in parentheses)

Seasons/ guilds	Number of Omnivorous Species	Number of Insectivorous Species	Number Frugivorous Species	Number of Carnivorous Species	Number of Granivorous Species	Number Nectarivorous Species
Monsoon 1988	10 (29.41)	10 (29.42)	7 (20.59)	4 (11.76)	1 (2.94)	2 (5.88)
Monsoon 1989	11 (47.83)	5 (21.74)	4 (17.39)	2 (8.70)	0 (0)	1 (4.35)
Monsoon 1990	12 (30.00)	15 (37.50)	10 (25.00)	3 (7.50)	0 (0)	0 (0)
Monsoon 1992	10 (30.30)	12 (36.36)	5 (15.15)	3 (9.09)	2 (6.06)	1 (3.03)
Summer 1989	14 (29.17)	14 (29.17)	14 (29.17)	3 (6.25)	1 (2.08)	2 (4.17)
Summer 1990	16 (32.66)	12(24.49)	17 (34.69)	1 (2.04)	2 (4.08)	1 (2.04)
Summer 1991	13(28.89)	10 (22.22)	13 (28.89)	7 (15.56)	1 (2.22)	1 (2.22)
Summer 1992	16 (32.00)	18(36.00)	13 (26.00)	2 (4.00)	0 (0.00)	1 (2.00)
Summer 1993	12 (36.36)	10 (30.31)	8 (24.24)	1 (3.03)	1 (3.03)	1 (3.03)
X^2	2.51	1.5	11.58**	0.04	0.14	0.11

Omnivorous birds were maximum in monsoon season and in summer followed by insectivorous and frugivorous birds. As in the case of Silent Valley more frugivorous birds were found during summer season. Significant difference was found in the number of frugivorous birds between summer and monsoon seasons at Mukkali (c^2 = 3.93; P = 0.05). Carnivorous birds were also high in abundance during summer (Table 4.22).

Table 4.22: Foraging guild composition in different seasons at Mukkali (Percentages in parentheses)

Seasons/ guilds	Number of Omnivorous species	Number of Insectivorous Species	Number of Frugivorous species	Number of Carnivorous Species	Number of Graminivorous Species	Number of Nectarivorous Species	Number of Piscivorous Species
Monsoon 1988	5 (29.41)	7 (41.18)	3 (17.65)	1 (5.88)	0 (0)	1 (5.88)	0 (0)
Monsoon 1989	12 (52.17)	9 (39.13)	2 (8.69)	0 (0)	0 (0)	0 (0)	0 (0)
Monsoon 1990	11 (36.67)	8 (26.67)	5 (16.66)	3 (10.00)	1 (3.33)	2 (6.67)	0 (0)
Monsoon 1992	24 (40.68)	19 (32.21)	5 (8.47)	5 (8.47)	3 (5.09)	2 (3.39)	1 (1.69)
Summer 1989	9 (29.03)	11 (35.48)	6 (19.36)	3 (9.68)	0 (0)	2 (6.45)	0 (0)
Summer 1990	6 (41.03)	12 (30.77)	4 (10.26)	2 (5.13)	3 (7.69)	1 (2.56)	1 (2.56)
Summer 1991	22 (36.67)	15 (25.00)	12 (20.00)	7 (11.66)	3 (5.00)	1 (1.67)	0 (0)
Summer 1992	21 (36.84)	22 (38.60)	6 (10.53)	3 (5.26)	2 (3.51)	2 (3.51)	1 (1.75)
Summer 1993	18 (43.90)	12 (29.27)	5 (12.20)	3 (7.32)	2 (4.88)	1 (2.44)	0 (0)
$\chi 2$	2.13	2.81	3.93*	1.50	1.33	0.11	0.33

Seasonal changes in availability of food

Availability or occurrence of species at a given place is determined by a number of factors and one important factor, which controls species richness is the availability of food. Any studies on species diversity will be meaningless without the knowledge on food productivity in the area. For this reason adequacy of animal and plant food were measured during the study period.

Plant food

The studies on fruiting and flowering indicated that an increase in flowering occurred during December to January months (Fig. 4.7).

At least five species of trees were in fruits every month. This suggests a steady availability of food during all seasons. No pattern was obtained in flushing of new leaves or falling down of yellow leaves. Emergence of green leaves was observed in almost all months. In the case of yellow leaves, June and July showed the highest (Fig. 4.8).

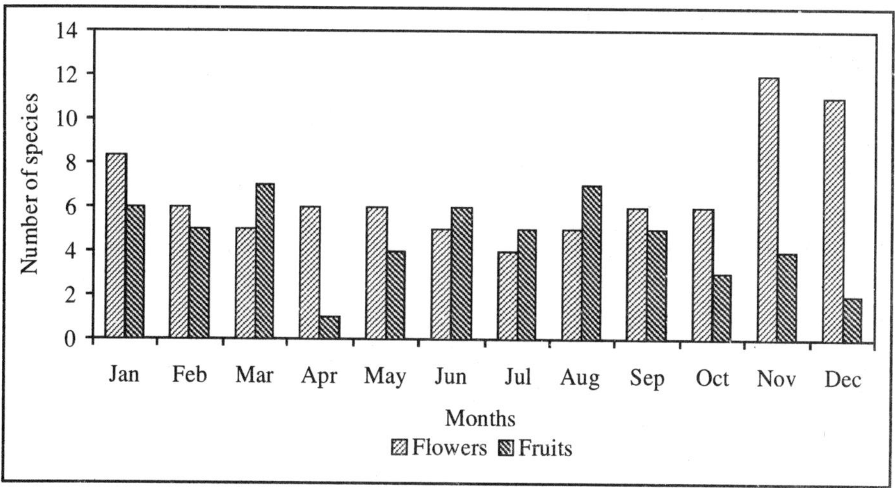

Fig. 4.7: Phenology of flowers and fruits at Silent Valley

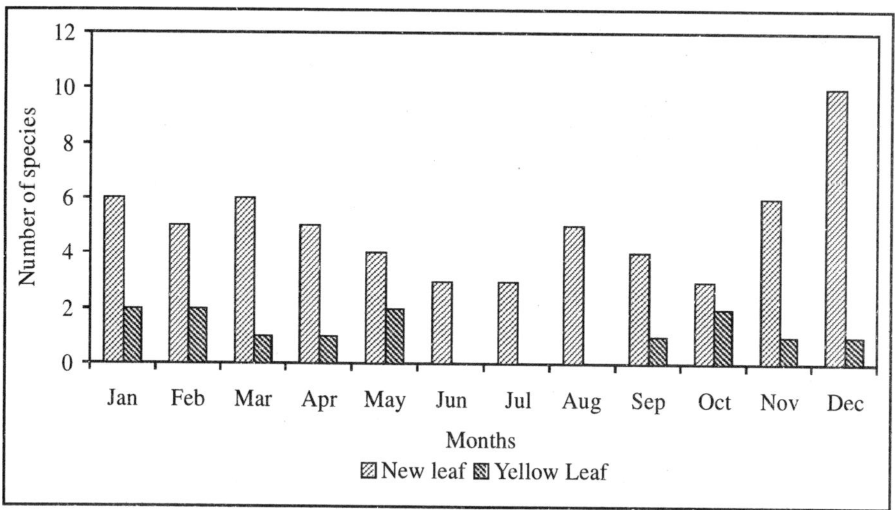

Fig. 4.8: Phenology of leaves at Silent Valley

No significant correlation is obtained between abundance of fruits and total number of birds in each month (r = 0.46; P = 0.05; n = 12). But a significant positive correlation is obtained between total number of birds and availability of flowers (r = 0.84; P = 0.001; n = 12; significant correlation obtained between new leaf emergence and total number of birds (r = 0.76; P = 0.01; n = 12).

Insect abundance

Insects were monitored every month to correlate the abundance of birds with that of the available food. All the insects collected were identified up to the Orders. They belonged to Orders like Orthoptera, Lepidoptera, Hymenoptera, Coleoptera, Hemiptera, Diptera, Odonata, Hymenoptera and Dictyoptera. Insect fauna showed two peaks of abundance. The first one was in the months of November to December and the second in the month of May. The first peak is after the southwest monsoon during November and December and the second with the first showers of southwest monsoon in May (Fig. 4.9).

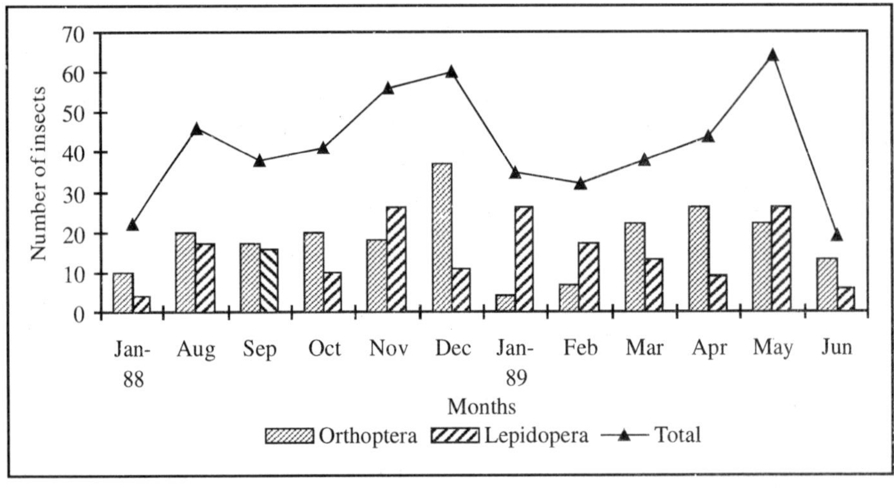

Fig. 4.9: Abundance of insects at Silent Valley

To find out the relationship between abundance of birds in different months and food availability the coefficient of correlation between the two events was calculated. No significant correlation obtained between the abundance of insects with the number of bird species in each month (r = 0.31; P = 0.05; n = 12) and density of birds (r = 0.54; P = 0.05; n = 12). However a significant correlation obtained between total number of birds and insect abundance (r = 0.60; P = 0.05; n = 12).

Monthly variation

Seasonal index showed that during the months of monsoon a reduction in the number of birds was observed at Silent Valley and Mukkali. Similar type of trends was reported from the tropical forests of other countries also. Variation in rainfall and soil moisture makes tropical bird fauna seasonal (Greenberg and Gradwohl, 1986). According to him this is due to the influence of rainfall on patterns of leaf, flower and fruit production, which in turn affect the population trends of arthropods.

Seasonal variation

Analysis on seasonal change in the abundance of birds also indicated the higher number of birds during summer season in the study area. Similar results were reported in an earlier study also (Jayson, 1989). When the two study sites are compared we can see that there was greater abundance of birds at Silent Valley during summer than at Mukkali. This can be attributed to the availability of more fruits at Silent Valley during summer. Density of birds and diversity indices were also higher at Silent Valley during summer. At Mukkali the bird population showed much more stability.

Factors influencing the bird community

Rainfall, temperature and availability of food appeared to be the factors as influencing the abundance of birds during different seasons in Silent Valley and Mukkali. Price (1979) who worked on the birds of Eastern Ghats found similar trend in annual cycles of bird fauna due to the seasonality of rainfall. As mentioned earlier, a few species of birds like the Yellowbrowed Bulbul showed stability in population even in the fluctuating environment. This can be attributed to the resident nature of the species coupled with its ability to feed on various food types like berries, drupes, nectar, spiders and insects.

At the Silent Valley the arthropod population showed an increase during the summer months soon after the rain with a corresponding increase in bird fauna. As only one method was employed for the collection of insects, information on insects in canopy levels is absent. According to Pyke (1984) the correspondence between the seasonal patterns of insectivore density and the abundance of flying insects is not very high and is not improved by focusing on birds that fed mostly on flying rather than non-flying insects. Even though sampling was concentrated on ground and herb level, an indication on the change in insect abundance in various months was provided by this collection method.

Foraging guild analysis showed a significant increase in the number of frugivorous birds during summer season at Silent Valley. Stiles (1978) also showed that even in tropical forest areas bird communities fluctuate in number in response to the availability of food and changes of climate. The relationship between food resources and bird diversity was also described by Terborgh (1985). Even though tropical forest birds are considered sedentary, in a study MacArthur (1972) has shown that seasonal movements are fundamental in many species's adaptive strategies in tropical forest habitats. The increase of bird population during summer may be attributed to the seasonal movements influenced by climate and availability of food.

❑❑❑

Chapter 5

Vertical Stratification of the Community

Foliage diversity of a forest stand has an important role in providing food for tropical forest birds. This feature adds to the ability of tropical forests to harbour high bird diversity. Studies on forest bird communities mainly examined the parameters like structure of forest bird communities (Nilson, 1983); distributions (Howe *et al.*, 1981) and community organisation (Landers and Mac Mohan, 1980). The pioneering, studies of Mac Arthur and Mac Arthur (1961) established the relationship between bird diversity and foliage density. Mac Arthur *et al.* (1962) and Mac Arthur *et al.* (1966) supported the above hypothesis. Some studies have showed negative relationship also (Wiens, 1983). Other vegetation characters, which have significant effect on bird diversity are foliage volume, percentage vegetation cover (Karr and Roth, 1971), percentage canopy cover (Crawford *et al.*, 1981; Wiens and Rottenberry, 1981; James and Wamer, 1982). In India Ramakrishnan (1983) examined many bird community parameters at Silent Valley. Diversity and structure of birds were also studied by Johnsingh *et al.* (1987), Katti (1989), Daniels (1989), Sundaramoorthy

(1991), Johnsingh and Joshua (1994), Gokula and Vijayan (1996). Even though many studies have been reported on the forest bird communities from South India, relation between foliage abundance and bird diversity have not been dealt in detail. The correlation between the abundance of birds in different height classes of tropical forest canopy with foliage abundance is reported in this chapter. The study forms part of a major investigation (Jayson, 1994), which determined many ecological aspects of two bird communities. Seasonal changes in these bird communities were reported earlier (Jayson and Mathew, 2000).

Analysis of vegetation

To analyse the foliage density at different canopy levels forest canopy at Silent Valley and Mukkali was divided into seven height classes. Data from both the vegetation types were pooled separately and classified into seven groups, namely 0 m, 1 - 5 m, 6 - 10 m, 11 - 15 m, 16 - 20 m, 21 - 30 m, and 31 ->31 m. From this, the preferred height for each species of birds was assessed. Species richness, abundance and diversity indices of birds in the each of the height classes were calculated, using the programme SPDIVERS.BAS (Ludwig and Reynolds, 1988). In order to correlate the vertical distribution of bird community and foliage abundance in different layers of canopy, the foliage abundance at different height levels were assessed. This was measured using the method described by Schemske and Brokaw (1981). Transects were laid on both sides of the main transect line (3 m away) at one m interval from each transect. Measurements were taken at one m interval on each transect line and thus 45 points were covered on each side, providing 90 points from both sides. Measurement was made at five-m height interval on each point. Height classes were 0 - 5 m, 5 - 10 m, 10 - 15 m, 15 - 2 m, 20 - 25 m, and 25 - >30 m. Foliage abundance at each height category was assessed by recording the number of touches by vegetation to a vertical pole at each strata. Data from 90 points were pooled and depicted for each point. Each such sites, 500 m apart, distributed in the main transect line were enumerated two times in two area in each season (summer and monsoon) for making foliage height profiles.

Birds

After considering various methods, variable width line transects was adopted for the census of birds (Burnham *et al.*, 1981). In this method the observer walks through a fixed path counting the birds seen or heard on both sides of the path. Whenever a bird was spotted it was identified up to species and details like the number of birds and perpendicular distance from transect were also noted (Ali, 1969; Ali and Ripley, 1983; Ali and Ripley, 1983a). The vertical distribution of bird community in the forest canopy was recorded along with the census of birds. Whenever a bird was sighted in the canopy, the height at which it was observed was also noted. This was used to analyse the vertical distribution pattern of birds in the foliage. Two line transects were selected one at Silent Valley and another one at Mukkali and each transect was 4 km in length. The first transect covered representative habitats of the area like evergreen forest, small patch of grassland, fire burned evergreen forest, and second transect covered the moist deciduous forest. Along the transect some rocky patches were also seen and some of the areas have a history of fire 10 years ago. Detailed census methodology was given in Jayson and Mathew (2000).

To demonstrate the relationship between foliage abundance and other parameters such as height, total number of birds, bird diversity and bird species richness, the values were compared using coefficient of correlation. For this mean of foliage abundance at each stratum was calculated from the eight samples in each area.

Species richness of birds in different height strata

Evergreen forests: Height preference of 94 species of birds was recorded from Silent Valley (Table 5.1). Only six species of birds used all the seven classes of height categories in the evergreen forests. These species can be termed as generalists and prominent among them were Black Bulbul (*Hypsipetes madagascariensis*) and Black Drongo (*Dicrurus adsimilis*) (Table 5.2). Six species were ground feeders using the heights between 0-5 meters. The vertical section up to 20 m was extensively utilised by a larger number of species than those above 20 m. Birds of prey like the Crested Serpent Eagle (*Spilornis cheela*) and the Blackwinged Kite (*Elanus caeruleus*) were seen always in the top canopy. The highest number of bird species (57) was recorded from the second height stratum starting from 1 to 5 m, followed by third (53) and fourth strata (45). The lowest number of species was seen in the top stratum. Species richness indices like Margalef Index showed the same trend. Highest value was recorded for the second height stratum (7.67) followed by third (7.48) and fourth strata (6.64). But highest value for Menhinick Index was obtained for the third (1.64) and fourth strata (1.64).

Table 5.1: Number of birds recorded in different height strata at Silent Valley

Sl. No.	Common name	Height (m)						
		0	1-5	6-10	11-15	16-20	21-30	31->31
1	Pond heron	1	-	-	1	-	-	-
2	Black-winged Kite	1	-	-	-	1	6	6
3	Brahminy Kite	-	1	-	-	-	-	1
4	Ceylon Shikra	1	1	-	-	-	-	1
5	Black Eagle	-	-	1	-	1	2	7
6	Crested Serpent Eagle	-	-	-	-	2	2	2
7	Shaheen Falcon	-	-	-	-	-	-	1
8	Painted Bush Quail	75	-	-	-	-	-	-
9	Grey Junglefowl	91	6	-	1	-	-	-
10	Red Spurfowl	4	-	-	-	-	-	-
11	Green Imperial Pigeon	-	-	2	-	-	-	-
12	Imperial Pigeon	-	2	1	2	-	6	7
13	Grey-fronted Green Pigeon	-	20	14	3	2	2	-
14	Yellow-legged Green Pigeon	-	3	31	12	-	3	-
15	Blue Rock Pigeon	2	-	-	-	-	-	-
16	Nilgiri Wood pigeon	-	7	-	-	-	-	-
17	Emerald Dove	12	11	3	-	-	1	-
18	Spotted Dove	-	4	5	-	-	-	-
19	Plum-headed Parakeet	2	1	6	23	4	-	16
20	Malabar Lorikeet	-	4	3	2	-	4	-
21	Blue-winged Parakeet	11	7	8	2	-	7	-
22	Rose-ringed Parakeet	12	18	26	59	10	22	-
23	Red-winged-crested Cuckoo	-	-	1	-	-	-	-
24	Short-eared Owl	-	-	-	1	-	-	-
25	White-rumped spinetail Swift	-	2	-	-	-	-	-
26	Malabar Trogon	-	1	-	-	-	-	-
27	Chest-nutheaded Bee-eater	-	-	1	1	1	-	-
28	Malabar Grey Hornbill	8	2	6	5	1	3	-
29	Great Indian Hornbill	-	-	5	3	-	1	2
30	Small Green Barbet	13	42	41	6	17	30	12
31	Golden-backed Woodpecker	-	11	23	16	10	13	7
32	Heart-spotted Woodpecker	-	5	2	-	-	-	-
33	Great-black Woodpecker	-	2	4	3	-	3	-
34	Threetoed Golden-backed Woodpecker	-	1	-	1	-	-	-
35	House Swallow	-	6	14	40	-	17	40
36	Red-rumped Swallow	-	-	-	-	-	11	-
37	Golden Oriole	1	-	3	2	-	-	-

Contd...

Table Contd...

38	Black-naped Oriole	-	-	-	-	-	1	-
39	Black-headed Oriole	-	-	-	1	-	-	-
40	Black Drongo	13	19	30	22	3	29	12
41	White-bellied Drongo	-	1	-	-	-	-	-
42	Bronzed Drongo	-	-	3	5	5	4	-
43	Racket-tailed Drongo	3	5	15	11	3	5	1
44	Common Myna	-	-	-	2	-	-	-
45	Hill Myna	-	11	37	101	48	160	63
46	Southern Tree Pie	13	8	31	23	9	2	-
47	Tree Pie	-	-	1	-	-	-	-
48	Golden-fronted Chloropsis	-	2	-	-	-	1	-
49	Scarlet Minivet	7	17	10	12	7	1	4
50	Common Iora	-	-	-	1	-	-	-
51	Fairy Bluebird	-	-	-	1	-	-	-
52	Ruby-throated Yellow bulbul	-	-	-	1	-	-	-
54	Red-whiskered Bulbul	31	89	45	11	-	1	1
55	Red-vented Bulbul	-	10	6	-	-	-	-
56	Yellow-browed Bulbul	57	405	380	138	47	30	12
57	Black bulbul	10	20	119	114	95	93	37
58	Jungle Babbler	102	112	10	8	-	-	-
59	Brown Shrike	-	-	1	-	1	-	-
60	White-bellied blue flycatcher	-	2	-	-	-	-	-
61	Scimitar Babbler	8	48	8	-	-	-	-
62	White-headed Babbler	-	7	-	-	-	-	-
63	Spotted Babbler	-	16	-	-	-	-	-
64	Rufous Babbler	-	4	-	-	-	-	-
65	Jungle Babbler	-	-	-	-	8	-	-
66	Brown Flycatcher	-	-	-	-	1	-	-
67	Brown-breasted Flycatcher	1	-	-	-	-	-	-
68	Tickell's Blue Flycatcher	-	7	-	-	-	-	-
69	Nilgiri Flycatcher	-	-	1	-	-	-	-
70	Paradise Flycatcher	2	4	1	1	-	-	-
71	Leaf Warbler	-	-	-	7	-	-	-
72	Magpie Robin	-	2	1	-	-	-	-
73	Pied Bush Chat	36	146	28	5	-	-	-
74	Malabar Whistling Thrush	31	24	18	11	-	-	-
75	Orange-headed Ground Thrush	-	-	-	1	-	-	-
76	White-throated Ground Thrush	5	15	10	-	-	-	-
77	Black Bird	1	-	-	-	-	1	-
78	Yellow-cheeked Tit	2	41	10	2	-	-	-
79	Grey Tit	-	-	3	-	-	-	-
80	Velvet-fronted Nuthatch	-	6	4	10	-	3	-
81	Indian Tree Pipit	-	-	1	-	-	1	-

Contd...

Table Contd...

82	Yellow Wagtail	28	6	1	-	-	-	-
83	Grey Wagtail	1	-	-	-	-	-	-
84	Tickell's Flowerpecker	-	3	1	-	-	-	-
85	Purple-rumped Sunbird	-	2	-	-	-	-	-
86	Loten's Sunbird	-	4	3	1	-	-	-
87	Little Spiderhunter	-	1	12	-	-	-	-
88	Small Sunbird	16	124	-	-	-	-	-
89	Nilgiri White-eye	6	47	36	10	-	-	-
90	Black-headed Munia	-	64	1	-	40	-	-
91	Rufous-bellied Munia	-	-	2	-	-	-	-
92	Spotted Munia	-	1	13	-	-	-	-
93	White-throated Munia	-	-	-	8	-	-	-
94	Common Rosefinch	4	61	-	6	-	8	-

- = (no birds recorded)

Table 5.2: Species richness of birds in different height strata at Silent Valley (Evergreen Forest)

Height strata	No. of species	Margalef Index	Menhinick Index
0 m	35	5.30	1.42
1 - 5 m	57	7.67	1.48
6 - 10 m	53	7.48	1.64
11 - 15 m	45	6.64	1.64
16 - 20 m	22	3.65	1.24
21 - 30 m	32	5.03	1.47
31 - >31 m	21	3.66	1.36

The total number of birds occupying in each height stratum varied at Silent Valley. The highest number of individuals (1,492) were observed in the second height stratum of 1 - 5 m (Fig. 5.1) and the next maximum was found in the third stratum (6-10 m). The height stratum above 30 m contained only few individuals.

Moist deciduous forests : Vertical height preference of 90 species was recorded from Mukkali (Table 5.3). At Mukkali also highest number of species was recorded from the second height stratum (68), followed by third (54) and fourth strata (44). Lowest species richness was obtained at the top most stratums. Highest value of Margalef Index was obtained again at the second stratum (9.71) followed by third (8.03) and fourth (7.03). Lowest value for both indices were recorded from the top most stratums (Table 5.4). Menhinick Index also showed a similar trend. At Mukkali the maximum number of birds were also found at the second stratum (991) followed by others in decreasing order (Fig. 5.2). However, one difference here was the occurrence of only few individuals at the top stratum, starting from 31 m and above. One disparity between the evergreen forest and moist deciduous forest was the paucity of birds above 30 m. In the moist deciduous forest the section up to 20 m was used extensively. A few parakeets, which were utilising the upper canopy, were the only occupants of the portion above 30 m. Only eight species were found to use all the first six height categories in the moist deciduous forests.

Table 5.3 : Number of birds recorded in different height strata at Mukkali

Sl. No	Species	Height class (m)						
		0	1-5	6-10	11-15	16-20	21-30	31->31
1	Blackwinged Kite	-	-	1	3	-	1	-
2.	Black Eagle	-	-	2	-	-	-	-
3.	Crested Serpent Eagle	1	1	1	1	-	1	-
4.	Falcon	-	-	-	1	-	-	-
5.	Ceylon Shikra	-	-	-	1	-	-	-
6.	Grey Junglefowl	24	1	-	1	-	-	-
7.	Grey-fronted Green Pigeon	-	-	8	-	3	-	-
8.	Spotted Dove	22	20	16	3	-	-	-
9.	Blossom-headed Parakeet	-	19	36	29	8	3	-
10.	Blue-winged Parakeet	-	-	4	8	-	1	-
11.	Rose-ringed Parakeet	1	5	25	21	12	3	11
12.	Malabar Lorikeet	-	4	5	4	-	0	-
13.	Brain fever Bird	-	-	1	-	-	-	-
14.	Indian Koel	-	-	-	-	1	-	-
15.	Shorteared Owl	-	-	-	1	-	-	-
16.	Crow Pheasant	-	1	1	-	-	-	-
17.	Jungle Owlet	-	1	-	-	1	-	-
18.	Malabar Trogon	-	2	-	-	-	-	-
19.	White-breasted Kingfisher	-	6	1	-	-	-	-
20.	Chest-nutheaded Bee-eater	-	4	4	-	1	-	-
21.	Hoopoe	2	-	-	-	-	-	-
22.	Malabar Grey Hornbill	-	-	-	3	2	1	-
23.	Great Indian Hornbill	-	-	1	-	-	-	-
24.	Small Green Barbet	8	24	53	43	15	9	-
25.	Small Yellownaped Woodpecker	-	2	2	-	1	-	1
26.	Goldenbacked Woodpecker	1	10	21	9	14	3	-
27.	Heartpotted Woodpecker	-	4	1	1	-	-	-

Contd...

Table Contd...

28.	Threetoed Goldenbacked Woodpecker	-	3	4	-	-	-	-
29.	Indian Pitta	3	2	-	-	-	-	-
30.	House Swallow	-	12	11	16	6	2	-
31.	Red-rumped Swallow	-	-	1	-	-	-	-
32.	Grey Shrike	-	1	-	-	-	-	-
33.	Bay-backed Shrike	-	6	1	1	1	-	-
34.	Rufous-backed Shrike	-	1	-	-	-	-	-
35.	Golden Oriole	-	-	-	1	-	-	-
36.	Black-headed Oriole	4	9	8	10	1	-	-
37.	Brown Shrike	-	5	-	-	-	-	-
38.	Black Drongo	5	47	53	46	9	4	1
39.	Racket-tailed Drongo	1	17	35	22	10	3	-
40.	Bronzed Drongo	1	3	1	10	-	4	-
41.	Hair-crested Drongo	-	1	-	-	-	-	-
42.	Common Myna	12	3	11	23	1	3	-
43.	Jungle Myna	2	2	-	2	-	1	-
44.	Hill Myna	-	-	3	9	4	9	-
45.	Southern Tree Pie	4	3	8	10	1	-	-
46.	Tree Pie	-	10	26	14	2	1	-
47.	Smaller Grey Cuckoo-Shrike	-	2	1	-	-	-	-
48.	Common Iora	-	1	5	-	-	-	-
49.	Scarlet Minivet	3	7	26	27	10	-	-
50.	Fairy Bluebird	-	2	-	-	1	-	-
51.	Golden-fronted Chloropsis	-	12	21	22	-	-	-
52.	Gold-mantled Chloropsis	-	3	3	2	2	2	-
53.	Yellow-browed Bulbul	-	35	38	10	2	2	-
55.	Ruby-throated Yellow Bulbul	-	1	1	-	-	-	-
56.	Black Bulbul	-	4	21	19	2	2	-
57.	Red-vented Bulbul	12	49	38	21	7	1	-
58.	Red-whiskered Bulbul	7	71	67	20	7	-	-
59.	Rufous Babbler	-	6	8	-	-	-	-
60.	Jungle Babbler	47	384	94	5	-	-	-
61.	Scimitar Babbler	-	-	2	-	-	-	-
62.	White-headed Babbler	6	35	1	-	-	-	-
63.	Black-and-Orange Flycatcher	-	-	-	-	-	1	-
64.	Brown-breasted Flycatcher	-	2	-	-	-	-	-
65.	Rufous-tailed Flycatcher	-	1	-	-	-	-	-
66.	Verditer Flycatcher	-	1	-	-	-	-	-
67.	Paradise Flycatcher	-	3	-	-	-	-	-
68.	Tailor Bird	-	4	1	-	-	-	-
69.	Leaf Warbler	-	2	-	-	-	-	-
70.	Pied Bush Chat	7	27	4	1	-	-	-
71.	Magpie-Robin	17	20	11	6	-	-	-
72.	Blue Rock Thrush	-	13	-	1	5	-	-

Contd...

Table Contd...

73.	White-throated Ground Thrush	5	11	1	1	-	-	-
74.	Orange-headed Ground Thrush	-	1	-	-	-	-	-
75.	Malabar Whistling Thrush	1	6	6	-	-	-	-
76.	Black Bird	-	-	5	7	-	-	-
77	Grey Tit	-	-	12	1	-	-	-
78.	Yellow-cheeked Tit	-	7	-	1	-	-	-
78.	Velvet-fronted Nuthatch	1	2	-	1	-	-	-
79.	Forest Wagtail	4	1	-	-	-	-	-
80.	Yellow Wagtail	22	-	-	-	-	-	-
81.	Pied Wagtail	1	-	-	-	-	-	-
82.	Tickell's Flowerpecker	2	-	-	-	-	-	-
83.	Loten's Sunbird	-	2	7	-	-	-	-
84.	Small Sunbird	-	2	-	-	-	-	-
85.	Purple-rumped Sunbird	-	4	-	-	-	-	-
86.	Little Spider Hunter	-	1	-	-	-	-	-
87.	Nilgiri White-Eye	-	30	4	16	-	-	-
88.	Yellow-throated Sparrow	-	1	-	-	-	-	-
89.	Black-headed Munia	19	2	11	-	-	-	-
90.	Rufous-bellied Munia	-	7	-	-	-	-	-

(- = no birds recorded)

Table 5.4: Species richness of birds in different height strata at Mukkali (Moist Deciduous Forest)

Height strata	No. of species	Margalef Index	Menhinick Index
0 m	30	5.27	1.92
1 - 5 m	68	9.71	2.16
6 - 10 m	54	8.03	1.99
11 - 15 m	44	7.03	2.07
16 - 20 m	27	5.35	2.38
21 - 30 m	21	4.95	2.78
31 - >31 m	3	0.77	0.83

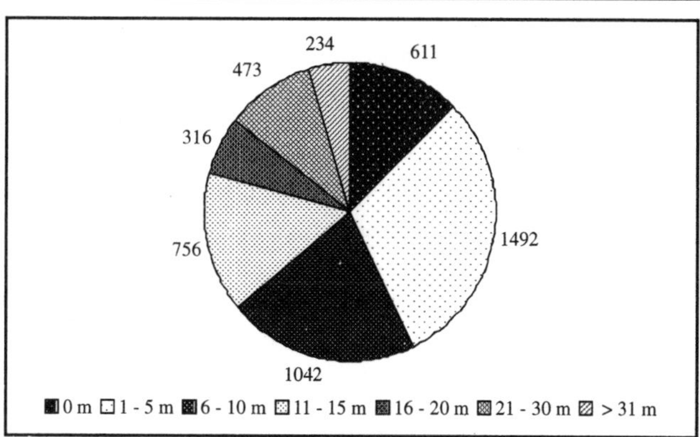

Fig. 5.1: Abundance of birds in each height strata (Silent Valley)

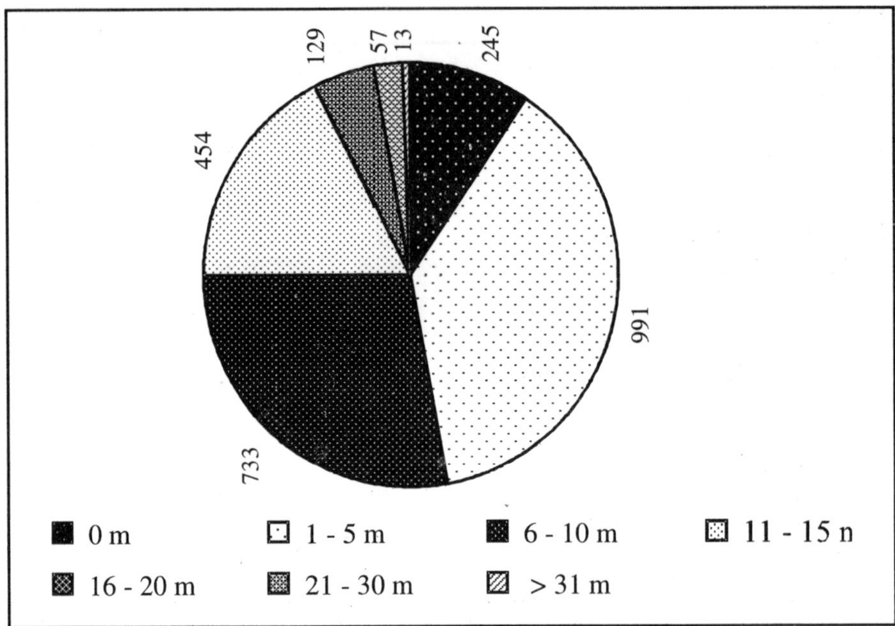

Fig. 5.2: Abundance of birds in each height strata (Mukkali)

Diversity of birds in different height strata

Evergreen forests: At Silent Valley the highest diversity index (H[1]) was obtained in the second stratum (1-5 m) followed by third (6-10 m) and fourth (11-15 m). A slight reduction of diversity was observed from the fifth stratum onwards. Lowest diversity was recorded at the fifth stratum (16-20 m). All the three indices showed the same pattern of decrease. Hill's Number N1 and N2 show high difference in the values (Table 5.5).

Table 5.5: Diversity Indices of birds in each height category at Silent Valley (Evergreen Forests)

Height Strata	Shannon-Wiener Index	Simpson's Index	Hill's Number N1	Hill's Number N2
0 m	2.79	0.09	16.25	11.36
1-5 m	2.89	0.11	17.96	9.29
6-10 m	2.71	0.16	15.05	6.35
11-15 m	2.80	0.09	16.49	10.59
16-20 m	2.23	0.16	9.30	6.39
21-30 m	2.35	0.17	10.51	5.91
31->31 m	2.35	0.14	10.50	7.40

Moist deciduous forests: Another pattern of difference in diversity index was obtained from Mukkali. The highest value of Shannon-Wiener index (3.28) was obtained in the middle canopy starting from 6-10 m of height and the lowest at the top most stratums above 31 m (0.54). All the three indices showed the same pattern. Simpson's index also showed the same trend. In Hill's Numbers N1 and N2 the difference in values are more obvious (Table 5.6).

Table 5.6: Diversity Indices of birds in each height category at Mukkali (Moist Deciduous Forests)

Height Strata	Shannon-Wiener Index	Simpson's Index	Hill's Number N1	Hill's Number N2
0 m	2.83	0.08	16.90	12.51
1-5 m	2.81	0.17	16.51	6.00
6-10 m	3.28	0.05	26.47	18.93
11-15 m	3.25	0.05	25.89	21.14
16-20 m	2.91	0.06	18.34	16.61
21-30 m	2.73	0.06	15.93	15.50
31->31 m	0.54	0.71	1.71	1.42

Foliage abundance in different height strata

In order to correlate the vertical distribution of bird community and foliage abundance in different layers of vegetation, the foliage abundance at various height levels were assessed as described in methods. Using this data two foliage profile diagrams were drawn for each vegetation type in summer and monsoon. At Silent Valley good foliage abundance was recorded up to 30 m height during monsoon (Fig. 5.3). In summer also similar observations were obtained. At Mukkali good foliage abundance was recorded up to 20-m height during summer and monsoon (Fig. 5.4).

Relation between foliage and abundance of birds

The results showed that as the height increased the foliage density decreased in both the vegetation types. The highest number of birds was also found in the strata of the highest foliage abundance. As the foliage abundance increased the diversity of birds and species richness were also increased at Silent Valley (Fig. 5.5). Similar patterns were found in the moist deciduous forest also (Fig. 5.6). Total number of birds decreased as the foliage abundance reduced. High diversity and species richness were also found in the strata of higher foliage density.

One difference found at Mukkali was that the decrease in the two upper most levels was higher than that at Silent Valley. At Silent Valley significant negative correlation was obtained between foliage abundance and height (r = - 0.84; P = 0.05; n = 6). Foliage abundance and number of individuals (r = 0.93; P = 0.01; n = 6) and foliage abundance and species richness (r =0.85; P =0.01; n = 6) also showed significant positive relationship. But no significant positive correlation obtained between bird diversity indices at each height stratum with foliage abundance (r = 0.79; P = 0.01; n = 6).

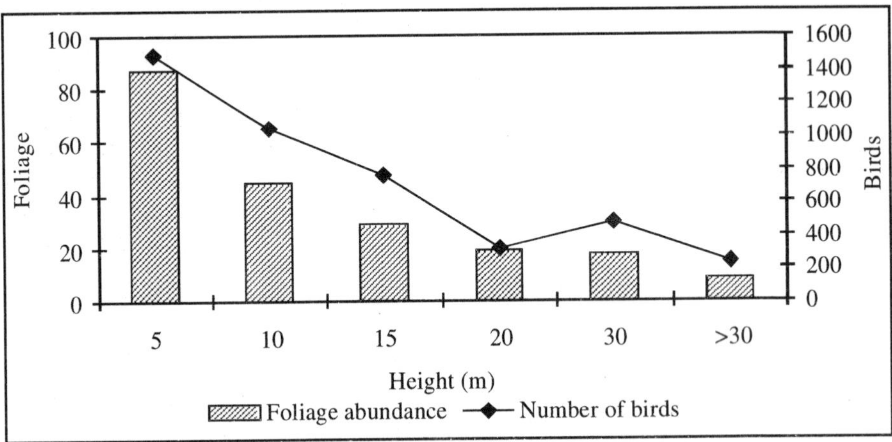

Fig. 5.3: Relation between foliage abundance, height and number of individuals in Evergreen forest (Silent Valley)

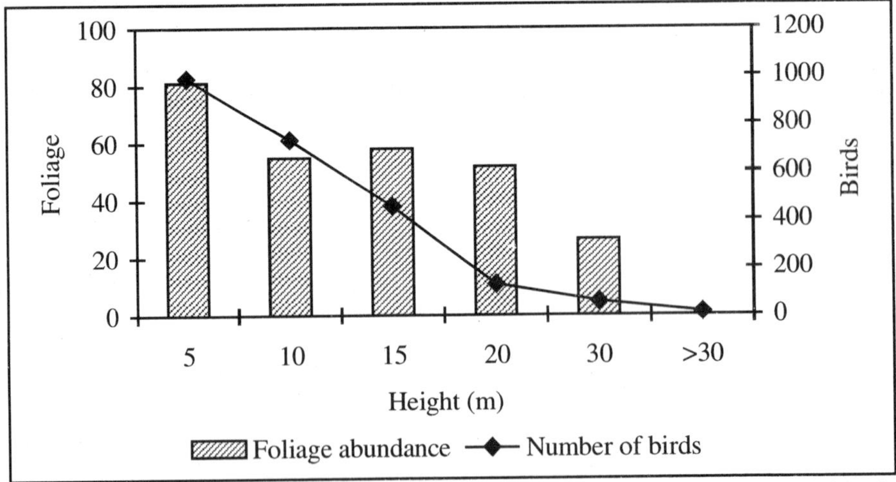

Fig. 5.4: Relation between foliage abundance, height and number of individuals in Moist deciduous forest (Mukkali)

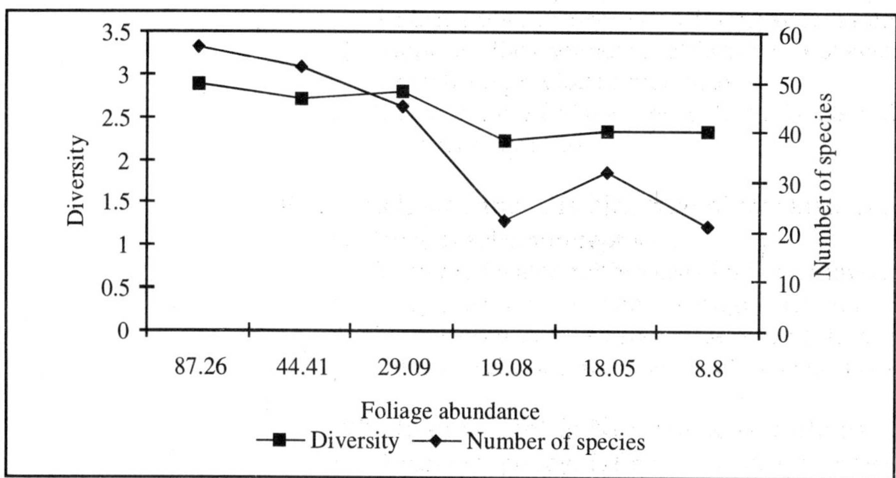

Fig. 5.5: Relation between foliage abundance, diversity and number of species in Evergreen forest (Silent Valley)

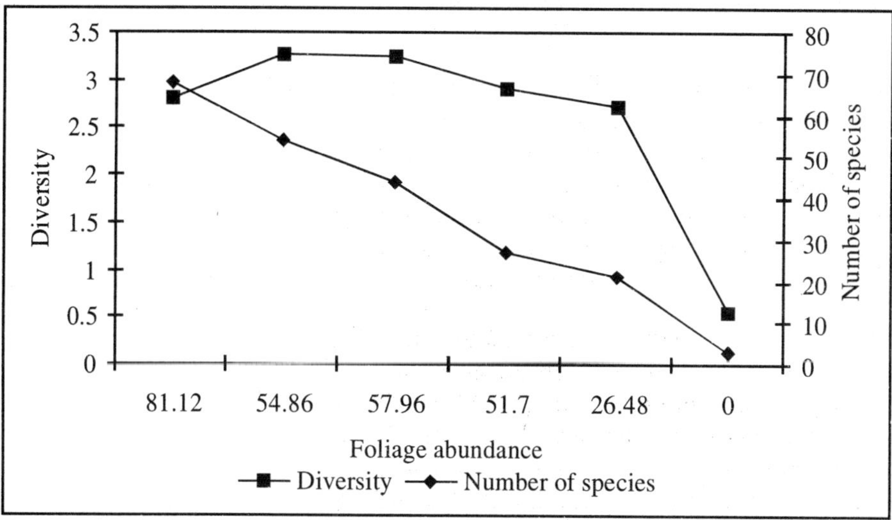

Fig. 5.6: Relation between foliage abundance, diversity and number of species in Moist deciduous forest (Mukkali)

At Mukkali also significant negative correlation was obtained between height and foliage (r = - 0.96; P = 0.01; n = 6). Here also the number of individuals and foliage abundance showed significant positive correlation (r = 0.84; P = 0.01; n= 6). Significant correlation was obtained between number of species and foliage abundance (r = 0.93; P = 0.01; n = 6). No significant positive correlation was obtained between foliage abundance and species diversity indices (r = 0.79; P = 0.01; n = 6).

Out of the 94 species of birds only six species has been using all the height categories. This indicates the preference of species to particular foliage availability. The usage may be related to the availability of food or cover. Species richness and diversity indices values showed that middle canopy are highly utilised by the birds than the bottom or top canopy. The study clearly showed the positive correlation between foliage abundance and species richness. Highest bird species richness and number of individuals were found in the height stratum where the highest foliage abundance was recorded.

As in the case of evergreen forest, highest species richness and species diversity were obtained in the middle canopy. An obvious positive co-relation was also obtained between foliage abundance and bird species richness and total number of birds. Relationship between vertical foliage diversity and bird diversity had been established in 1960's. But until Mac Arthur and Mac Arthur (1961) devised a quantitative measurement of vertical foliage diversity it was not possible to depict this phenomenon quantitatively. He obtained foliage diversity by visually estimating the proportion of total foliage in chosen horizontal layers and obtained best relationships in three horizontal layers in (0 to 0.7 m), (0.7 to 7.6 m) and (7.6 to >7.6 m) in temperate Central America. For Australia Recher (1969) and for Europe Moss (1978) proved this relationship. Other workers also demonstrated the relationship between bird diversity and vegetation structure (Cody, 1970; Adeline, 1984 and Verner and Larson, 1989). Pearson (1982) could not find any relationship between vegetation structure and community.

Ramakrishnan (1983) recorded the diversity of bird fauna in different height strata, but not the relationship between vertical foliage diversity and bird diversity. This study indicates significant relationship between foliage abundance and other bird community parameters in a tropical forest. Other vegetation characters, which have significant effect on bird diversity, are foliage volume, percentage vegetation cover (Karr and Roth, 1971), percentage canopy cover (Crawford *et al.*, 1981: Wiens and Rottenberry, 1981: James and Wamer, 1982). Ground feeders like Grey jungle fowl, babblers and ground thrushes contributed to the higher number of omnivorous species at ground level. Insectivorous birds that depended on foliage were more in the middle canopy. Occurrence of nectar feeding birds in the lower canopy is explainable to the abundance of flowers in that level produced by the herbs and shrubs.

This study indicated significant positive relationship between foliage abundance and species richness and also total number of birds in the evergreen and moist deciduous forests of Silent Valley and Mukkali. Vertical distribution of bird species in the canopy is related to microhabitat use and resource availability (Wiens, 1989). In this study no attempt has been made to assess the availability of food resources in each height strata. As this study suggested a significant positive correlation between foliage abundance and abundance of birds, future studies can attempt to quantify resource availability in each height strata.

□□□

Chapter 6

Habitat Utilisation

In this chapter utilisation of habitat by the birds of the Silent Valley National Park has been analysed. Usually three questions are asked in the habitat utilisation studies (1) whether the species form well defined assemblages, quite apart from the geographical distribution patterns? (2) Whether ecologically or taxonomically similar species segregate by habitat? (3) Whether habitat characteristics attribute for the variations in the distribution or abundance (Wiens, 1989).

METHODS

Whenever a bird was observed in the period of study, its exact location and activities were recorded. The concept of habitat as described by Odum (1974), "the habitat of an organism is the place where it lives or the place where one would go to find it" is adopted for this study. Four main habitats were identified in the study area, depending on the vegetation type namely, Evergreen Forest (EF), Moist Deciduous Forest (MDF), Coffee Estate (CE) and Agricultural Area (AA). In each of the above habitats, "microhabitat" of birds was also recorded (Odum, 1974). Microhabitats identified at Silent Valley were the tree trunk, foliage, branch, bushes, dead tree, ground, grass and open area. Microhabitats identified at Mukkali were tree trunk, foliage, branch, bushes, ground, rocky areas and open area.

Niche breadth

To find out the overlap in foraging parameters Niche breadth (B), Levins Standardised Niche breadth (BA) and Niche overlap indices were calculated (Ludwig and Reynolds, 1988). Data were collected on 58 more frequently observed species of birds from the study area. Among this fifty-eight species, six were congeners; they were Parakeets *Psittacula* sp. (2), Drongos *Dicrurus* sp. (3), Myna *Acridotheres* sp. (2), Tree Pie *Dendrocitta* sp. (2), Bulbuls *Pycnonotus* sp. (3) and Bulbuls *Hypsipetes* sp. (2).

Habitat use

Those species consisting at least 2 per cent and above of the whole data only were considered for analysis. Quantitative data for 21 species is given in Table 6.1. Spotted Doves were sighted in agricultural areas and coffee estate without much difference. Rose-ringed Parakeets were mainly sighted in the evergreen forest. But most of the Blossom-headed Parakeets were sighted in moist deciduous forest. The Malabar Lorikeets were using branches of tree and foliage in all habitats.

The Small Green Barbet was basically using tree branches, bushes and foliage, both in evergreen forest and in moist deciduous forest. Among the woodpeckers, Golden-backed Woodpecker was found to use dry tree branches, and tree trunk in both the areas. The Heart-spotted Woodpeckers were using smaller branches. Among the orioles, the Black-headed Oriole was using tree branches and foliage at both the areas. The Common Myna, which was recorded only from Mukkali was using moist deciduous forest, estate and agriculture areas. In the same way Jungle Myna was using burned areas and moist deciduous forest. Hill Mynas were found mainly in evergreen forests of Silent Valley. Among the Tree Pies, the Southern Tree Pie was found in the evergreen forests and moist deciduous forests. But the Tree Pie was found only in the moist deciduous forests of Mukkali.

Table 6.1: Number of birds found in each habitat in the study area

S. No.	Common name	Habitats				(n)
		EF	MDF	CE	AA	
1.	Spotted Dove	1	3	7	12	23
2.	Rose-ringed Parakeet	13	4	6	3	26
3.	Blossom-headed Parakeet	4	14	8	-	26
4.	Malabar Lorikeet	5	7	4	10	26
5.	Small Green Barbet	21	17	16	2	56
6.	Golden-backed Woodpecker	13	11	6	-	30
7.	Heart-spotted Woodpecker	14	13	-	-	27
8.	Black Drongo	21	11	28	1	61
9.	Racket-tailed Drongo	5	10	8	1	24
10.	Hill Myna	40	5	-	-	45
11.	Southern Tree Pie	15	5	2	-	33
12.	Red-whiskered Bulbul	14	3	6	15	38
13.	Yellow-browed Bulbul	110	23	10	-	143
14.	Red-vented Bulbul	14	5	9	5	33
15.	Black Bulbul	20	3	-	-	23
16.	Jungle Babbler	12	19	21	9	61
17.	Malabar Whistling Thrush	16	3	9	-	28
18.	Yellow-cheeked Tit	30	5	-	-	35
19.	Loten's Sunbird	12	8	2	-	22
20.	Nilgiri White-eye	20	4	3	-	27
21.	Magpie-robin	-	18	14	10	42

EF - Evergreen Forest, MDF - Moist deciduous Forest,
CE - Coffee Estate, AA - Agriculture area

Common Iora was found only in the moist deciduous forest of Mukkali. The Golden-fronted Chloropsis was maximum recorded in the moist deciduous forest of Mukkali. Bulbuls were found in all habitats at Silent Valley and Mukkali. Red-whiskered Bulbuls used rocky areas, bushes and burnt bushes, moist deciduous forest, estate and agriculture areas. Yellowbrowed Bulbul was highest at evergreen forest of Silent Valley. The Redvented Bulbul was numerous in the moist deciduous forests of Mukkali, and in the estate. The Rubythroated Yellow Bulbul was recorded only from moist deciduous forest. The Black Bulbul was highest in the evergreen forest followed by moist deciduous forest. The Jungle Babbler used evergreen forest, moist deciduous forest, and estate and agriculture areas. The Whiteheaded Babbler used moist deciduous forest and estate. Magpie-robin was seen only in the moist deciduous forest, agricultural area and in the estate.

The Malabar Whistling Thrush was observed in evergreen forests, estates, and moist deciduous forests. The Yellow cheeked Tit utilised bushes in the moist deciduous forest, and Velvet fronted Nuthatch evergreen forests, estate and moist deciduous forest. Yellow Wagtail used both evergreen forest and moist deciduous forest. The Grey Wagtail was found only in evergreen forest. Similarly Pied Wagtail was found only at Mukkali. Sunbirds were utilising both evergreen forest and moist deciduous forest and Little Spiderhunter both evergreen forest and moist deciduous forests. Nilgiri White-eyes occurred in both evergreen and moist deciduous forest. The White-throated Munia was found only at Silent Valley. But Rufousbellied Munia and Black-headed Munia were found in evergreen forests and moist deciduous forests. Common Rosefinches were found only in evergreen forests.

Since the Silent Valley transect covered degraded and burned evergreen forests, some species of birds, which are found in other forest types were also recorded from here Black Drongo, Jungle Babbler and Rose-ringed Parakeet.

Microhabitat use

Microhabitat use of 20 species was recorded from the forests of Silent Valley. Similarly microhabitat use of 16 species was recorded from moist deciduous forests, 17 species from coffee estate and 11 species from agriculture area at Mukkali. The details are given in the Tables 6.2, 6.4 and 6.6.

Table 6.2: Microhabitat use of birds in the forests of Silent Valley (percentages)

Sl. No.	Species	Microhabitats								Niche Breadth		
		Foliage	Tree branch	Tree trunk	Bushes	Ground	Dead tree branch	Rocky area	Open area	n	B	BA
1.	Grey Junglefowl	-	-	-	14	86	-	-	-	14	1.31	0.04
2.	Emerald Dove	65	10	-	-	25	-	-	-	20	2.02	0.15
3.	Roseringed Parakeet	-	62	-	-	-	15	-	-	13	2.17	0.17
4.	Chestnutheaded Bee-eater	-	-	-	-	-	92	-	8	12	1.17	0.02
5.	Small Green Barbet	9	67	-	10	-	14	-	-	21	2.05	0.15
6.	Goldenbacked Woodpecker	8	8	69	-	-	15	-	-	13	1.96	0.14
7.	Heartpotted Woodpecker	-	-	57	-	-	43	-	-	14	1.96	0.14
8.	Black Drongo	9	24	-	-	-	43	5	-	21	3.95	0.42
9.	Hill Myna	-	35	2	-	-	50	-	13	40	2.56	0.22
10.	Southern Tree Pie	33	40	-	27	-	-	-	-	15	2.93	0.28
11.	Redwhiskered Bulbul	14	14	-	36	7	7	22	-	14	4.41	0.49
12.	Yellowbrowed Bulbul	44	23	-	30	1	1	-	2	110	2.97	0.28

Contd...

Table Contd...

13.	Redvented Bulbul	-	50	-	50	-	-	-	-	14	2.00	0.14
14.	Black Bulbul	5	45	-	5	-	45	-	-	20	2.44	0.21
15.	Jungle Babbler	-	-	-	25	69	-	6	-	12	1.84	0.12
16.	Malabar Whistling Thrush	25	12	-	-	63	-	-	-	16	2.11	0.16
17.	Yellowcheeked Tit	17	17	50	16	-	-	-	-	30	2.30	0.29
18.	Yellow Wagtail	-	-	-	-	92	-	-	-	13	1.17	0.02
19.	Loten's Sunbird	-	-	-	-	-	42	-	58.3	12	1.95	0.14
20.	Nilgiri White-eye	50	25	-	25	-	-	-	-	20	2.67	0.24

n = Sample size

Niche breadth and overlap

Highest niche breadth was for Red-whiskered Bulbul followed by Black Drongo. Niche overlap indices computed are given below.

GO	GOmin	GOadj	V	df
0.425	0.067	0.384	745.37	133

The general overlap (GO) between 20 species in the microhabitat use is low. As the transformed value of V = 22.33 which is greater than the tabled value of standard normal distribution, 1.96 (P = 0.05; df = 133), it can be concluded that there is no complete overlap in microhabitat use. Specific Overlap (SO) analysis showed, high overlap between the following pairs of species (P = 0.05; df = 7). All other species showed no high overlap (Table 6.3).

Table 6.3: Pairs of species showing high specific overlap (SO) values in microhabitat use at Silent Valley

Species	SO	U
Grey Junglefowl and Yellow Wagtail	0.889	2.827
Small Green Barbet and Black Bulbul	0.802	9.276
Grey Junglefowl and Jungle Babbler	0.583	12.936
Grey Junglefowl and Malabar Whistling Thrush	0.625	11.280
Malabar Whistling Thrush and Emerald Dove	0.696	11.574
Chestnutheaded Bee-eater and Hill Myna	0.593	12.524
Chestnutheaded Bee-eater and Loten's Sunbird	0.571	13.454
Black Bulbul and Small Green Barbet	0.760	11.001
Goldenbacked Woodpecker and Heartpotted Woodpecker	0.719	9.224

Niche breadth and overlap (MDF)

Highest niche breadth was for Common Myna. Niche overlap indices are given below.

GO	GOmin	GOadj	V	df
0.430	0.066	0.390	347.564	120

Table 6.4: Microhabitat use of birds in the moist deciduous forest of Mukkali (Percentages)

Sl. No.	Species	Microhabitats									n	Niche B	Breadth BA
		Foliage	Tree branch	Tree trunk	Bushes	Ground	Dead tree branch	Rocky area	Open area	Burned area			
1.	Blossomheaded Parakeet	14	86	-	-	-	-	-	-	-	14	-1.32	0.04
2.	Malabar Lorikeet	57	43	-	-	-	-	-	-	-	7	-1.96	0.14
3.	Small Green Barbet	-	71	29	-	-	-	-	-	-	17	-1.70	0.10
4.	Goldenbacked Woodpecker	-	55	-	-	-	45	-	-	-	11	-1.98	0.14
5.	Heartspotted Woodpecker	-	62	38	-	-	-	-	-	-	13	-1.89	0.13
6.	Indian Pitta	-	28	-	29	43	-	-	-	-	14	-2.88	0.27
7.	Black Drongo	36	64	-	-	-	-	-	-	-	11	-1.85	0.12
8.	Racket-tailed Drongo	10	70	-	-	-	-	-	20	-	10	-1.85	0.12
9.	Common Myna	-	50	25	-	8	8	-	9	-	12	-2.99	0.25
10.	Jungle Myna	-	-	25	-	50	-	-	-	25	12	-2.67	0.21
11.	Goldenfronted Chloropsis	50	50	-	-	-	-	-	-	-	8	-2.00	0.13
12.	Yellowbrowed Bulbul	44	56	-	-	-	-	-	-	-	23	-1.97	0.12
13.	Jungle Babbler	-	16	-	63	21	-	-	-	-	19	-2.14	0.14
14.	Magpie-robin	-	28	-	-	72	-	-	-	-	18	-1.67	0.08
15.	Yellow Wagtail	-	-	-	-	89	-	11	-	-	9	-1.24	0.03
16.	Loten's Sunbird	12	88	-	-	-	-	-	-	-	8	-1.27	0.03

The general overlap (GO) between 16 species in their microhabitat use in the moist deciduous forest of Mukkali was low. When we compare the transformed value of V (V = 10.94) with the critical value of standard normal distribution = 1.96 (P = 0.05; df = 120) we can conclude that the General overlap is far from complete. Specific overlap (SO) was high between the 37 pairs of species (P = 0.05; df = 8). All other species showed only low overlap. Nine pairs showing higher values are listed below (Table 6.5).

Table 6.5: Pairs of species showing high specific overlap (SO) in microhabitat use in the moist deciduous forests of Mukkali

Species	SO	V
Malabar Lorikeet and Black Drongo	0.915	1.244
Malabar Lorikeet and Golden-fronted Chloropsis	0.990	0.143
Malabar Lorikeet and Yellow-browed Bulbul	0.963	0.526
Small Green Barbet and Heart-spotted Woodpecker	0.982	0.610
Small Green Barbet and Common Myna	0.747	9.901
Black Drongo and Yellow-browed Bulbul	0.996	0.230
Black Drongo and Golden-fronted Chloropsis	0.963	0.829
Chloropsis and Yellow-browed Bulbul	0.991	0.137
Small Green Barbet and Common Myna	0.747	9.901

Niche breadth and overlap

Highest niche breadth was recorded for Red vented Bulbul. General overlap (GO), minimum GO (GO min) and adjusted GO (GO adj) are given below.

GO	GOmin	GOadj	V	df
0.396	0.068	0.352	318.319	112

The General overlap between 17 species in their microhabitat use in the estate of Mukkali is low. As the transformed value of V (10.33) is greater than the critical value of standard normal distribution = 1.96 (P = 0.05; df = 112), we can assume that there is no complete overlap between the species. Specific overlap (SO) showed complete overlap between 27 pairs of species (P = 0.05, df = 7). Most of the species were utilizing branches of trees followed by foliage. But in the coffee estate the second position was for bushes. All other microhabitats have only limited use. Branches of dead trees and trunks of trees were highly used at Silent Valley. But as dead trees were low at Mukkali it was not available for birds (Table 6.7).

Table 6.6: Microhabitat use of birds in the coffee estate of Mukkali (Percentages)

Sl. No.	Species	Microhabitats									n	Niche B	Breadth BA
		Foliage	Tree branch	Tree trunk	Bushes	Ground	Dead tree branch	Rocky area	Open area	Burned area			
1.	Blossomheaded Parakeet	14	86	-	-	-	-	-	-	-	14	-1.32	0.04
2.	Malabar Lorikeet	57	43	-	-	-	-	-	-	-	7	-1.96	0.14
3.	Small Green Barbet	-	71	29	-	-	-	-	-	-	17	-1.70	0.10
4.	Goldenbacked Woodpecker	-	55	-	-	-	45	-	-	-	11	-1.98	0.14
5.	Heartpotted Woodpecker	-	62	38	-	-	-	-	-	-	13	-1.89	0.13
6.	Indian Pitta	-	28	-	29	43	-	-	-	-	14	-2.88	0.27
7.	Black Drongo	36	64	-	-	-	-	-	-	-	11	-1.85	0.12
8.	Racket-tailed Drongo	10	70	-	-	-	-	-	20	-	10	-1.85	0.12
9.	Common Myna	-	50	25	-	8	8	-	9	-	12	-2.99	0.25
10.	Jungle Myna	-	-	25	-	50	-	-	-	25	12	-2.67	0.21
11.	Goldenfronted Chloropsis	50	50	-	-	-	-	-	-	-	8	-2.00	0.13
12.	Yellowbrowed Bulbul	44	56	-	-	-	-	-	-	-	23	-1.97	0.12
13.	Jungle Babbler	-	16	-	63	21	-	-	-	-	19	-2.14	0.14
14.	Magpie-robin	-	28	-	-	72	-	-	-	-	18	-1.67	0.08
15.	Yellow Wagtail	-	-	-	-	89	-	11	-	-	9	-1.24	0.03
16.	Loten's Sunbird	12	88	-	-	-	-	-	-	-	8	-1.27	0.03

n = Sample size

Table 6.7: Microhabitats and number of species using them in various habitats

S. No.	Microhabitats	Silent Valley Evergreen forest	Mukkali Moist deciduous forest	Coffee estate
1.	Foliage	11 (17.19)	7 (17.95)	6 (12.77)
2.	Tree branch	14 (21.88)	14 (35.90)	15 (31.91)
3.	Bushes	10 (15.63)	2 (5.13)	7 (14.89)
4.	Dead tree branch	11 (17.19)	2 (5.13)	5 (10.64)
5.	Tree trunk	4 (6.25)	4 (10.26)	6 (12.77)
6.	Rocky area	3 (4.69)	1 (2.56)	—
7.	Ground	7 (10.94)	6 (15.38)	4 (8.51)
8.	Burned area	—	1 (2.00)	—
9.	Open area	4 (6.25)	2 (5.13)	4 (8.51)

Percentages in parenthesis

The results show that at Silent Valley four Species of birds were using only one microhabitat, seven species using two microhabitats and only one species was found to use more than 8 microhabitat (Fig. 6.1). This showed the limited diversity of habitats at Silent Valley. In the case of Mukkali only 15 species were using one microhabitat while 9 species were found to use more than 8 microhabitats.

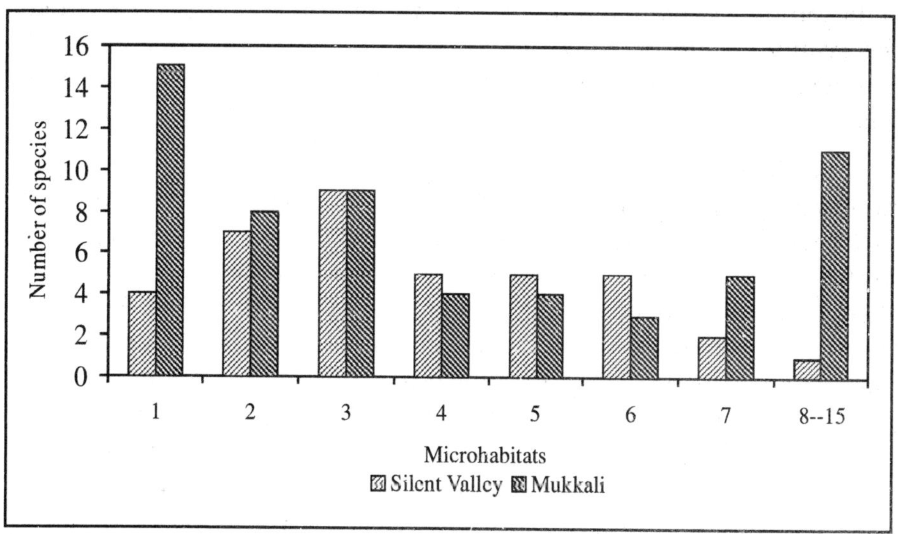

Fig. 6.1: Microhabitat use of different species in the study area

Many factors are known to influence the decisions made by the individual birds in selecting a place in which to establish a breeding territory or to escape from a predator. Habitat or microhabitat is not used only for feeding, but birds use them for nesting, courtship and mating, singing, resting (Wiens, 1989). Microhabitat use may vary over time and between locations. Table 6.5 indicate that the species like Yellowbrowed Bulbul, Hill Myna, Southern Tree Pie, Black Bulbul and Nilgiri White-eye were more numerous in the evergreen forests of Silent Valley. As wetland habitat was rare at the study site waterfowls and wetland birds were not common. Similarly Magpie-robin, Jungle Babbler, and Black Drongo was more in the moist deciduous forest.

Silent Valley

Among the eight microhabitats identified at Silent Valley most of the species were found among the foliage. The general overlap in microhabitat use between the species is low. Only 9 pairs showed high overlap in microhabitat use.

Mukkali

In the moist deciduous forests of Mukkali most of the birds were depending on the branches of trees rather than foliage. Even though general overlap (GO) shows similar values as that of Silent Valley, specific overlap (SO) analysis shows high overlap in 37 pairs of species. Less specilization in habitat use by the birds at Mukkali can be attributed to the simple vegetation structure. From the coffee estates also similar observations were obtained. Here the value of general overlap (GO) was very low and 27 pairs of birds show high overlap in microhabitat use.

Specilization in habitat use among the congeneric species is worth mentioning. Among them Grey-fronted Green Pigeon was found on the tree branches, while Yellow-legged Green Pigeon was found mostly on the ground. Black Drongos were found in the open areas while its related species Bronzed Drongo was mainly on tree branches. Southern Tree Pie was mostly found in evergreen forest of Silent Valley and were absent in moist deciduous forests of Mukkali. Red-vented Bulbuls were rare in the evergreen forests of Silent Valley, while it was abundant in Mukkali. At Mukkali 51 species were (49 per cent) using moist deciduous areas, where as only 33 (31 per cent) species were using estates and 22 (20 per cent) species were using agricultural areas. Most of the species at Silent Valley were depending on tree branches

(21.88 per cent) followed by foliage (17.19 per cent). Since Silent Valley area is having tropical wet evergreen forest this can be expected.

At Mukkali also majority of species were utilizing branches of trees (35.90 per cent) and foliage (17.95 per cent) followed by other microhabitats. Same was the pattern in the estate. In the case of agricultural areas branches of trees and bushes were used with same degree. Here also no complete overlap was found in the habitat use. It is evident from the data that habitat characteristics are attributing to the variations in the distribution of taxonomically similar species such as Hill Myna and Common Myna, Jungle Babbler and Scimitar Babbler, Tree Pie and Southern Tree Pie, and Red-vented Bulbul, Yellow-browed Bulbul and Redwhiskered Bulbul. These species were segregated by their difference in their microhabitat use.

❑❑❑

PART - II

WETLAND BIRD COMMUNITIES

Chapter 7

Species Composition of Wetland Birds

Information on the avifauna of an area is a prerequisite to assess the status of birds and the habitat quality with specific attention to indicator species including the rare, endangered and endemic species. Birds are one of the best indicators of the health of an ecosystem. They are highly mobile and easily observed indicators of change in the environment (Holmes *et al.*, 1986). Many wetland species also play a role in the control of agricultural pests, while some species are themselves considered pest of certain crops. After fish, birds are probably the most important faunal group that attract people to wetlands. Loss of wetland habitats through direct and indirect modifications and non-sustainable harvesting of water birds for human needs have led to decline in several water bird populations and a number of species (Jin-Han Im *et al.*, 2001).

Some of the most catastrophic declines of birds have taken place in the last few decades and the list of threatened species in the Asia-Pacific region has expanded rapidly to include species from a large range of water bird groups. While decline of some populations has been well documented, the fates of many others remain unknown. It is vital to understand the underlying causes in population changes and to attempt to control these trends in order to prevent key component of biodiversity of wetland habitat from being lost. The number of water birds using a particular habitat is related to types and quality of habitats, abundance and availability of food and level of disturbance. Monitoring of water birds can thus provide valuable information on the status of wetlands and can be a key tool for increasing the awareness of importance of wetland and conservation values. In this chapter, the status, occurrence and species composition of avifauna recorded from Kole wetlands are elucidated.

Methods

Species richness and species composition of birds were computed from the data obtained through daily census and field observations. Birds were classified as migratory and resident species based on the occurrence data and published literature. Globally threatened species of birds were identified based on BirdLife International (2001). Feeding and guild composition were collected from the available literature (Ali and Ripley, 1983). Bird species have been categorised as aquatic feeders, insectivores, granivores, nectar-frugivores, carnivores, frugivores and omnivores. They were also classified as water birds, waders and terrestrial birds based on their habitat use.

Occurrence of species

One hundred and eighty two taxa of birds were recorded from the Kole wetlands. These belong to 50 Families under 16 Orders. Of the 182 species, 100 species were resident, 81 migrants and one species was a straggler. Out of these, 48 species were new records for the area. The highest number of species was recorded from Kanjany (121) followed by Parappur (117), Enamavu (94) and Chettupuzha (71). Little Egret (*Egretta garzetta*), Cattle Egret (*Bubulcus ibis*), Little Cormorant (*Phalacrocorax niger*), Whiskered Tern (*Chlidonias hybridus*), and Indian Pond-Heron (*Ardeola grayii*) were the most abundant species in the Kole wetlands.

Classification according to the habitat revealed that 25 per cent of birds were water birds followed by waders (21 per cent) and terrestrial birds (54 per cent), which showed that Kole wetlands is an abode of many passerine species.

Migratory species

Eighty one species of birds were migrants and among these 49 were trans-continental migrants and thirty two species were local migrants (Table 7.1). Migrants were classified as abundant, common and uncommon based on the number of individuals.

Table 7.1: Trans-continental migratory species of birds recorded from the Kole wetlands

Sl.No.	Common name	Scientific name	Abundance*
1.	Masked Booby	*Sula dactylatra*	U
2.	European White Stork	*Ciconia ciconia*	U
3.	Black Stork	*Ciconia nigra*	U
4.	Common Teal	*Anas crecca*	A
5.	Northern Pintail	*Anas acuta*	A
6.	Gadwall	*Anas strepera*	C
7.	Garganey	*Anas querquedula*	A
8.	Northern Shoveller	*Anas clypeata*	U
9.	Ferruginous Pochard	*Aythya nyroca*	U
10.	Pallid Harrier	*Circus macrourus*	U
11.	Pied Harrier	*Circus melanoleucos*	U
12.	Western Marsh-Harrier	*Circus aeruginosus*	C
13.	Eurasian Sparrowhawk	*Accipiter nisus*	U
14.	Watercock	*Gallicrex cinerea*	U
15.	Pied Avocet	*Recurvirostra avosetta*	U
16.	Pacific Golden-Plover	*Pluvialis fulva*	C
17.	Lesser Sand Plover	*Charadrius mongolus*	A
18.	Little Ringed Plover	*Charadrius dubius*	A
19.	Kentish Plover	*Charadrius alexandrinus*	U
20.	Common Snipe	*Gallinago gallinago*	U
21.	Pintail Snipe	*Gallinago stenura*	U
22.	Black-tailed Godwit	*Limosa limosa*	C
23.	Bar-tailed Godwit	*Limosa lapponica*	U
24.	Whimbrel	*Numenius phaeopus*	U
25.	Eurasian Curlew	*Numenius arquata*	C
26.	Common Redshank	*Tringa totanus*	U
27.	Marsh Sandpiper	*Tringa stagnatilis*	A
28.	Common Greenshank	*Tringa nebularia*	C
29.	Green Sandpiper	*Tringa ochropus*	C
30.	Wood Sandpiper	*Tringa glareola*	A
31.	Terek Sandpiper	*Xenus cinereus*	U
32.	Common Sandpiper	*Actitis hypoleucos*	A
33.	Ruddy Turnstone	*Arenaria interpres*	U
34.	Great Knot	*Calidris tenuirostris*	U

Contd...

Table Contd...

35.	Dunlin	*Calidris alpina*	C
36.	Broad-billed Sandpiper	*Limicola falcinellus*	C
37.	Curlew Sandpiper	*Calidris ferruginea*	A
38.	Sanderling	*Calidris alba*	C
39.	Little Stint	*Calidris minuta*	A
40.	Temminck's Stint	*Calidris temminckii*	C
41.	Ruff	*Philomachus pugnax*	A
42.	Yellow-legged Gull	*Larus cachinnans*	U
43.	Brown-headed Gull	*Larus brunnicephalus*	A
44.	Black-headed Gull	*Larus ridibundus*	A
45.	Caspian Tern	*Sterna caspia*	C
46.	Desert Wheatear	*Oenanthe deserti*	U
47.	Eurasian Tree pipit	*Anthus trivialis*	U
48.	Citrine Wagtail	*Motacilla citreola*	U
49.	Grey Wagtail	*Motacilla cinerea*	U

* Classification of birds is based on the numbers of individuals observed
A = Abundant (more than 1000 individuals)
C = Common (more than 100 - 1000 individuals)
U = Uncommon (less than 100 individuals)

Waders

Waders constitute an important group of wetland species. These birds depend heavily on shallow waters and mud flats, normally recorded from September onwards in the Kole wetlands. Details on the occurrence of waders in the four intensive study sites are presented in Table 7.2. The highest species of waders was recorded from Kanjany (31) followed by Enamavu (30), Parappur (18) and Chettupuzha (15). Black-winged Stilt (*Himantopus himantopus*), Little Ringed Plover (*Charadrius dubius*), Wood Sandpiper (*Tringa glareola*), Red-wattled Lapwing (*Vanellus indicus*), Common Greenshank (*Tringa nebularia*), Green Sandpiper (*Tringa ochropus*), Common Sandpiper (*Actitis hypoleucos*), and Whiskered Tern (*Chlidonias hybridus*) were recorded from all the intensive study sites in the three migratory seasons. Species like Great Knot (*Calidris tenuirostris*), Pied Avocet (*Recurvirostra avosetta*), Pintail Snipe (*Gallinago stenura*) and Caspian Tern (*Sterna caspica*) were recorded only from Kanjany and Dunlin (*Calidris alpina*) and Whimbrel (*Numenius phaeopus*) from Enamavu.

Table 7.2: Waders recorded from the intensive study areas in the Kole wetlands (1998-2001)

Sl. No.	Common name	Chettu-puzha	Kanjany	Enamavu	Parappur
1.	Pheasant-tailed Jacana	√		√	
2.	Bronze-winged Jacana	√	√	√	
3.	Greater Painted-Snipe		√	√	
4.	Pied Avocet		√		
5.	Small Pratincole		√	√	
6.	Black-winged Stilt	√	√	√	√
7.	Red-wattled Lapwing	√	√	√	√
8.	Pacific Golden-Plover	√	√		
9.	Lesser Sand Plover	√	√	√	
10.	Little Ringed Plover	√	√	√	√
11.	Kentish Plover		√	√	√
12.	Black-tailed Godwit	√	√	√	
13.	Bar-tailed Godwit		√	√	
14.	Whimbrel			√	
15.	Eurasian Curlew	√	√	√	
16.	Common Redshank		√	√	√
17.	Marsh Sandpiper	√	√	√	√
18.	Common Greenshank	√	√	√	√
19.	Green Sandpiper	√	√	√	√
20.	Wood Sandpiper	√	√	√	√
21.	Common Sandpiper	√	√	√	√
22.	Ruddy Turnstone			√	
23.	Great Knot		√		
24.	Dunlin			√	
25.	Curlew Sandpiper		√	√	√
26.	Broad-billed Sandpiper		√	√	√
27.	Common Snipe		√	√	√
28.	Pintail Snipe		√		
29.	Terek Sandpiper		√	√	
30.	Sanderling		√		√
31.	Little Stint		√	√	√
32.	Temminck's Stint		√	√	√
33.	Ruff		√	√	√
34.	Eurasian Woodcock			√	–
35.	Whiskered Tern	√	√	√	√
36.	Caspian Tern		√		
	Total	**15**	**31**	**30**	**18**

Order wise classification of avian species observed in the Kole wetlands is given in Table 7.3. The Order Passeriformes had the highest percentage (31) of species followed by Charadriiformes (21) and Ciconiiformes (12). Feeding guild analysis showed that majority of species were insectivores (51) followed by Omnivores (43) and aquatic feeders (42) (Table 7.3).

Table 7.3: Order wise classification and feeding guild composition of bird species recorded from the Kole wetlands

Sl. No.	Order	Status			Feeding guilds						
		R	M	Total	A	I	G	N/F	C	F	O
1.	Podicipidiformes	01	--	01	1	--	--	--	--	--	--
2.	Pelecaniformes	02	05	07	7	--	--	--	--	--	--
3.	Ciconiiformes	15	06	21	20	1	--	--	--	--	--
4.	Anseriformes	01	08	09	9	--	--	--	--	--	--
5.	Falconif ormes	05	05	10	--	--	--	--	10	--	--
6.	Galliformes	03	--	03	--	--	3	--	--	--	--
7.	Gruiformes	05	02	07	--	3	3	--	--	--	1
8.	Charadriiformes	03	36	39	--	2	--	--	--	--	37
9.	Columbiformes	03	--	03	--	--	3	--	--	--	--
10.	Psittaciformes	01	01	02	--	--	--	--	--	2	--
11.	Cuculiformes	03	03	06	--	--	--	--	--	6	--
12.	Strigiformes	03	--	03	--	--	--	--	3	--	--
13.	Apodiformes	03	--	03	--	3	--	--	--	--	--
14.	Coraciiformes	07	02	09	5	4	--	--	--	--	--
15.	Piciformes	02	--	02	--	2	--	--	--	--	--
16.	Passeri formes	43	14	57	--	36	9	5	--	2	5
	Total	**100**	**82**	**182**	**42**	**51**	**18**	**5**	**13**	**10**	**43**

R = Resident, M = Migrant; A = Aquatic feeders, I = Insectivores, G = Granivores, N/F = Nectar-Frugivores, C = Carnivores, F = Frugivores, O = Omnivores

A comparison of number of wetland bird species recorded from the Kole wetlands with those from Kerala, India, Asia and World is given in Table 7.4. Out of the 245 species of wetland birds recorded from India, 33 per cent were found in the Kole wetlands.

Table 7.4: Comparative occurrence of wetland bird species in the Kole wetlands

Sl. No.	Order and Family	World[1]	Asia[1]	India[2]	Kerala[3]	Kole Wetlands*
1.	Gaviiformes					
	Gavidae	5	4	2	-	-
2.	Podicipediformes					
	Podicipedidae	21	6	5	2	1
3.	Pelecaniformes					
	Pelecanidae	8	4	3	1	1
	Phalacrococidae	36	11	5	4	4
4.	Ciconiiformes					
	Ardeidae	60	30	24	15	13
	Ciconiidae	19	11	9	6	5
	Threskiornithidae	31	9	4	4	3
	Phoenicopteridae	5	2	2	1	-
5.	Anseriformes					
	Anatidae	149	62	45	12	9
6.	Gruiformes					
	Gruidae	15	9	7	-	-
	Rallidae	124	34	23	9	7
	Heliornithidae	3	1	1	-	-
7.	Charadriiformes					
	Jacanidae	8	3	2	2	2
	Haematopodidae	11	2	1	1	-
	Charadriidae	152	82	63	37	28
	Rostratulidae	2	1	1	1	1
	Recurvirostridae	12	3	3	2	2
	Dromadidae	1	1	1	1	-
	Burhinidae	9	3	4	2	-
	Glareolidae	18	6	6	3	1
	Laridae	81	45	35	21	5
	Total	**770**	**329**	**245**	**124**	**82**

1 - Alfred *et al.*, 2000; 2 - Ali and Ripley, 1983; 3 - Neelakantan *et al.*, 1993 and Ali, 1969; * - present study

Endemic and globally threatened species

Out of the 16 species of birds, which are endemic to the Western Ghats (Jhunjhunwala *et al.*, 2001), Indian Rufous Babbler (*Turdoides subrufus*) was recorded from the Kole wetlands. According to BirdLife International (2001), one hundred and twenty nine threatened bird species occur in India, of these six species were recorded from the Kole wetlands (Table 7.6).

Table 7.6: Globally threatened bird species recorded from the Kole wetlands

Sl. No.	Species	Status	Number of birds observed
1.	Oriental White Ibis	NT	1230
2.	Darter	NT	117
3.	Ferruginous Pochard	NT	82
4.	Pallid Harrier	NT	07
5.	Painted Stork	NT	05
6.	Spot-billed Pelican	VU	04

VU = Vulnerable; NT = Near Threatened

The number of species recorded from the Kole wetlands showed high species richness, which is comparable to other wetlands in Kerala. Before the present study was initiated, only 158 species of birds were reported from the area. Among these, fifty were wetland birds and four raptors (Anon., 1992 and 1993). During this study, 48 species were added to the existing list. Certain Passerine species, which were recorded earlier, were not spotted.

In the present study, 49 species of trans-continental migrants were recorded, which showed the importance of the area as a wintering ground for migratory species. Among the trans-continental migrants, Bar-tailed Godwit, Great Knot and Whimbrel are apparently capable of long distance flights (Driscoll and Ueta, 2000). Nine species of waterfowls were recorded from the Kole wetlands and among them, Gargany, Pintail, Lesser Whistling Teal and Common Teal were abundant. Rest of the species *viz.* Gadwall, Spot-billed Duck, Cotton Teal, Ferruginous Pochard and Northern Shoveller were sighted in few numbers *i.e.*, less than 500. The sighting of Black Stork from the Kole wetlands was the first record of the species from the coastal plains of Kerala. Ali (1969) has not reported Black Stork from Kerala, but later the species was recorded from Periyar Tiger Reserve, Thekkady by Kurup (1989). Previous records of Black Stork were from Parambikulam and Walayar in Palakkad District and Chamaravattam in Malappuram District (Neelakantan *et al.*, 1993).

Reports on Lesser Frigate Bird were very few from Kerala. Ferguson (1904) recorded Lesser Frigate Bird from Trivandrum and Ali (1969) has not reported the species from Kerala during his survey. Faizi (1985) reported the species from Quilon based on a museum specimen. The sighting of Lesser Frigate Bird during this study is the first report from Central Kerala (Jayson and Sivaperuman, 2003a). Similarly, Northern Shoveller was newly recorded during the study (Sivaperuman and Jayson, 2002a). As this wetland is coming under 'Central Asian - Indian flyway', protection of the migratory species is of highest priority. Kole wetland is an ideal habitat for migratory and resident birds, especially for the winter visitors.

❑❑❑

Chapter 8

Species - Abundance Relations

Number of individuals in a population increase under favourable environmental conditions. Sudden or violent changes in the environment or in the biological characteristics of the individual could lead to decline. Fluctuation in the population is usually because of the effect of a combination of several factors responsible for their survival, growth and reproduction. It is necessary to understand the cause of such fluctuations for better scientific management of the population as well as the habitat (Acharya, 2000). In most of the studies, community has largely been expressed in terms of species richness, abundance, density and diversity. All these have been used as indicators of habitat quality, because an increase in the value of the component is generally thought to reflect larger amounts of the necessary resources to sustain larger population within a given area. Moreover, it is believed that if the factors determining the distribution of the animals are known, predictions can be made concerning the responses of animals to specific perturbations and hence, such animals can be used to monitor environment quality (Gokula, 1998).

Information on the population of wetland birds in Kerala is scanty, except for a few surveys and reports. The population studies have traditionally been used to monitor large scale, long term changes in avian population and to assess both habitat quality and the responses of birds to both management practices and natural and human caused environmental changes (Wiens, 1989). Species-abundance relationships of whole bird community and wetland birds are described in this chapter.

METHODS

The study was conducted from November 1998 to October 2001 and it was mainly based on direct observational methods (Altman, 1974). The whole area was surveyed on foot and vehicle and all the important areas were visited. After the reconnaissance survey, four intensive study areas *viz.* Chettupuzha, Kanjany, Enamavu and Parappur were selected. These areas were selected since all the microhabitats were available and represented the main part of the Kole wetlands.

Census methods

After considering all available census methods, total count method was selected and bird population was estimated using the total count method (Hoves and Blakewell, 1989). Total count is an easy and accurate method for estimating the number of birds present in an area. It can be applied for large flock, densely packed flocks or distant flocks either in flight or on the ground. This method involves counting or estimating a "block" of birds within a flock. Depending on overall flock size, a "block" can be of 10, 100 or 1000 birds and then use these "block" as a model, to measure the remainder of the flock (Fig. 8.1).

In this method, representative blocks were identified and birds in the blocks were counted using a telescope (15x - 45x). These blocks had natural boundaries in the form of bunds. The time of observation was from 0700 hour to 1000 hour. On an average, 20 days were spent in the field in a month and no census was made during heavy rain. In the each intensive study area, four blocks of 10 ha were selected for bird census. Four counts were made in each block in each month and the pooled average was used for estimating species richness and abundance. Night surveys were also conducted for recording the nocturnal birds in the intensive study areas. Birds were identified based on physical features with the help of field guides and reference books (All and Ripley, 1983; Neelakantan, *et al.*, 1993: Grimmett *et al.*, 1998). During the sampling, all the birds seen were counted and the habitats of birds were also recorded. Based on the arrival of migratory birds, two different seasons were classified *viz.* migratory season (September to March) and non-migratory season (April to August).

Water depth in all the intensive study areas were measured using a nylon rope tied with a weight of 1 kg and lowered down till it touched. Measurements were recorded from different locations and four such measurements were taken and noted. The depth is expressed in centimeter. Rainfall and temperature data were collected from the Irrigation Department, Thrissur.

Community parameters like species richness, abundance, diversity indices, evenness indices, density of birds and similarity indices were calculated and presented for all the months. These parameters were also presented for each month and for each intensive study area by taking the mean values.

Collector's curve

The collector's curve is used for assessing the sampling efficiency. This curve was drawn by plotting the cumulative number of species observed in each month against the month of observation (Pielou, 1975).

Species richness and abundance

Species richness (number of species) and abundance (number of individuals) of birds in every month in the study area were calculated from the census data and field observations. Species richness indices like Margalef Index (R1) and Menhinick Index (R2) were calculated using the formula given by Magurran (1988). These indices provide easily understandable measures of diversity. Species richness as a yardstick of diversity was used in many earlier studies.

Margalef Index, $D_{mg} = (S-1) / 1n (N)$

[$\ln = \log_e$]

Menhinick Index, $D_{mn} = S/(N)^{**} \frac{1}{2}$

where

S = Number of species

N = Total number of individuals summed over all species

Fig. 8.1: Schematic representation of total count method Shaded circle = number of birds accurately counted Unshaded circles = number of birds estimated

Species diversity indices

Diversity measures the variation in richness and abundance. Diversity Index combines the information on multiple species into a single number. These indices provide easily understandable measures of diversity. Shannon-Weiner (H'), Simpson's (λ), and Hill's diversity number N1 and N2 were calculated using the computer program SPDIVERS.BAS developed by Ludwig and Reynolds (1988).

Shannon-Weiner Index (H')

Shannon-Weiner Index of general diversity (H') is given as

$$H' = - \Sigma P_i \log P_i$$

where Pi the proportional abundance of the ith species = (ni/N)

Simpson's Index (λ)

The following equation is used to calculate the Simpson's Index (λ)

$$\lambda = \Sigma (ni (ni-1)/(N (N-1))$$

where

ni = the number of individuals in the ith species

N = total number of individuals

Hill's diversity

Hill's diversity N1 is calculated from Shannon-Weiner Index

$$N1 = e^{H'}$$

In addition, Hill's diversity N2 is calculated from Simpson's Index

$$N2 = 1/lambda$$

Evenness measures

A number of indices have been used to quantify the evenness of diversity. Two evenness measures *viz.* Shannon Evenness and Sheldon Evenness were calculated using the computer program SPDIVERS.BAS developed by Ludwig and Reynolds (1988). The following formulae were used for calculating two Evenness measures based on Shannon-Weiner Index and Simpson's Index.

$$\text{Shannon Evenness (E1)} = \frac{H'}{\log(S)}$$

where

H′ = Shannon-Weiner Index

S = Number of species

$$\text{Sheldon Evenness (E2)} = \frac{e^{H'}}{S}$$

Density

Density of birds in each month and individual density of selected species were calculated for the whole area and for the intensive study areas. The density was estimated from the daily census of birds.

$$\text{Density (D)} = \frac{\text{Number of birds sighted}}{\text{Area surveyed (ha)}}$$

Similarity Indices

Similarity Indices between the intensive study areas were calculated using Jaccard Index and Sorenson Index (Magurran, 1988).

Jaccard Index

$$C_j = j / (a + b - j)$$

where

j = the number of species common to both sites

a = the number of species in site A and

b = the number of species in site B

Sorenson Index

$$C_S = 2j / (a + b)$$

where

j = the number of species common to both sites

a = the number of species in site A and

b = the number of species in site B

Distribution models

Species-abundance model was constructed as explained in Magurran (1988). Species of birds were ranked in order of abundance, as represented by individuals seen for each species and this was plotted in decreasing order for all species against the number of individuals for the whole area. Truncated lognormal distribution was fitted to species-abundance data, using maximum likelihood estimation (Slocomb *et al.*, 1977).

Dominance Index

The dominance of the each bird species in the Kole wetlands was calculated using the dominance index.

$$\text{Dominance Index} = n_i \times 100 / N$$

where

n_i = Number of individuals

N = Total number of all the species seen during the study period

Feeding guilds

Details of food habit of various species were collected from the literature (Ali and Ripley, 1983). Species richness and abundance of birds in different feeding guilds were also assessed. Bird species have been categorized into aquatic feeders, insectivores, granivores, nectar-frugivores, carnivores, frugivores and omnivores.

Collector's curve

The collector's curve showed the sampling efficiency of the study. An increase in the number of species was recorded from the month of November 1998 to January 2001. After that, there was no change in the species richness, which indicated that the sampling was adequate (Fig. 8.2).

Species richness

Species richness of birds varied in different months and the highest of 97 species was recorded during December 1999 and the lowest of 15 during June 1999 (Table 8.1). The overall variation in the species richness during the study period is presented in Fig. 8.3. Species richness increased during the migratory season and decreased during the southwest monsoon.

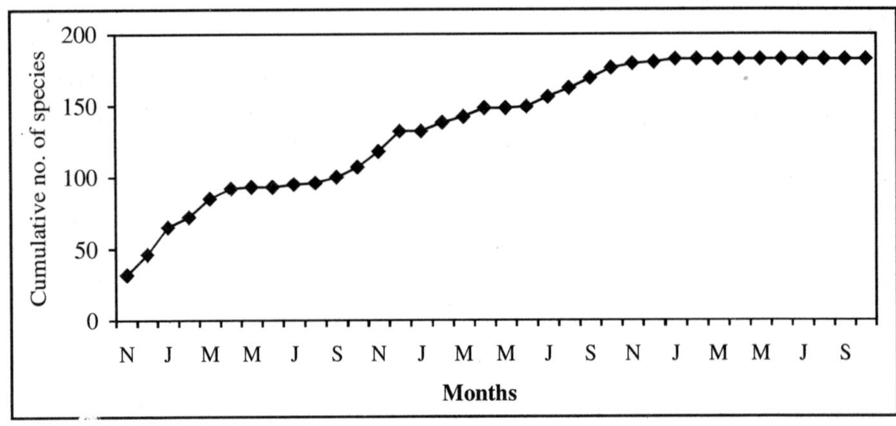

Fig. 8.2: Collector's curve from the Kole wetlands

Table 8.1: Species richness of birds in the Kole wetlands in different months (1998-2001)

Years	Months											
	J	F	M	A	M	J	J	A	S	O	N	D
1998	—	—	—	—	—	—	—	—	—	—	32	34
1999	49	46	51	40	33	15	23	25	38	71	61	97
2000	49	39	64	43	49	39	31	34	76	56	70	67
2001	55	56	64	61	19	26	31	31	34	49	—	—

— = no data collected

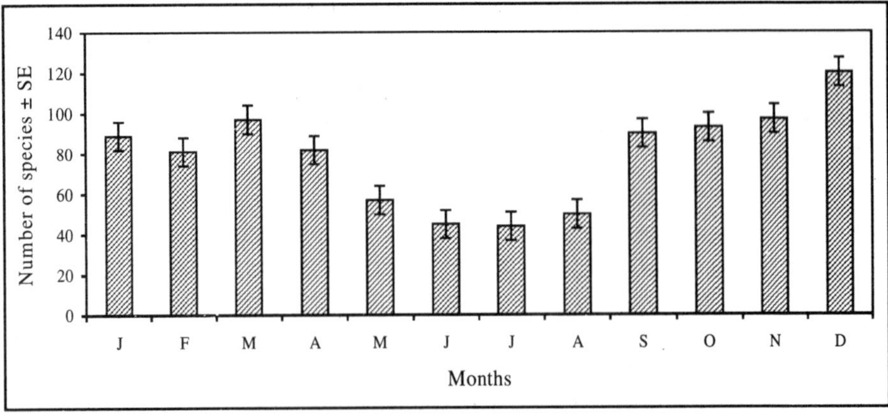

Fig. 8.3: Species richness of birds in the Kole wetlands (1998-2001)

Abundance of birds

Total number of birds varied from 35 to 8,033 individuals in a month. The highest number of birds (8,033) was recorded during November 2000 and the lowest (35) during June 1999 (Table 8.2). However, when all the months are taken together, the highest number of birds was observed during December and the lowest in July. As in the case of species richness, total number of birds was also low during the southwest monsoon (Fig. 8.4).

Table 8.2: Total number of birds recorded in the Kole wetlands in different months (1998-2001)

Years	Months											
	J	F	M	A	M	J	J	A	S	O	N	D
1998	—	—	—	—	—	—	—	—	—	—	414	891
1999	690	918	378	348	150	35	72	147	355	3071	7860	5843
2000	2721	918	1403	1200	1232	265	97	294	2772	2725	8033	3934
2001	1705	5266	5072	6211	1196	531	214	421	2105	1233	—	—

— = no data collected

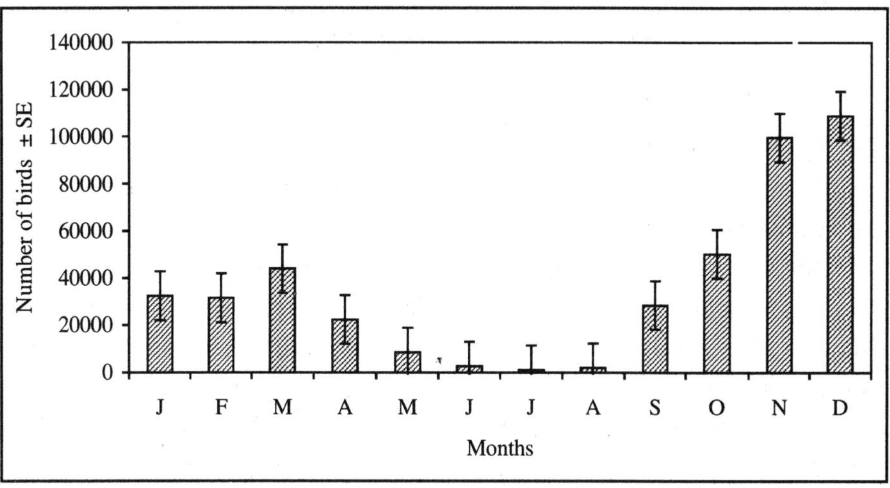

Fig. 8.4: Total number of birds recorded in the Kole wetlands in different months (1998-2001)

Diversity indices

Indices based on the proportional abundance of species are the best approach to measure diversity. Most widely used diversity indices like Shannon Index of diversity, Simpson's Index of diversity and Hill's numbers N1 and N2 have been determined. During the period of study, the highest diversity index (H') was recorded in the month of December (3.01), July (2.96) and the lowest H' value (2.11) was in October (Fig. 8.5).

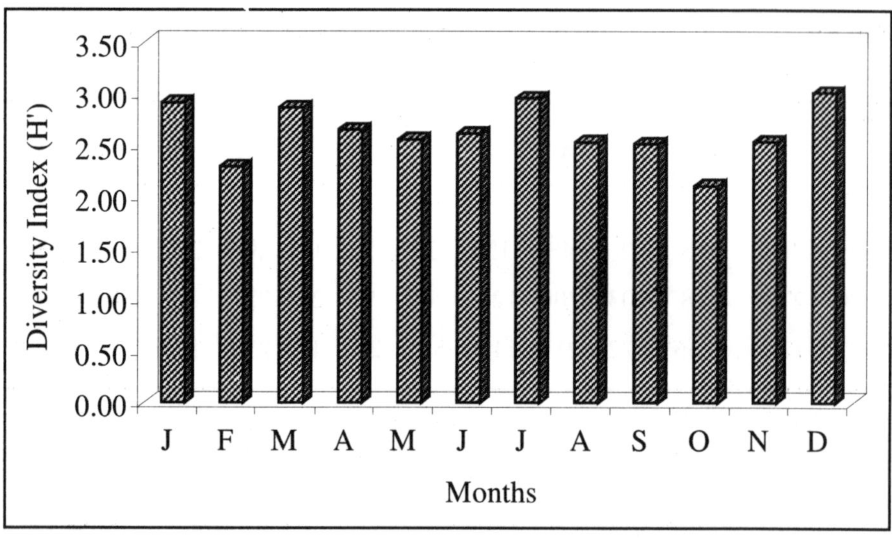

Fig. 8.5: Diversity index (H') in different months

Evenness indices

Evenness indices measure the evenness of species-abundance and are complimentary to the diversity index concept. It is a measure of how the individuals are appropriated among the species. Two of the evenness measures are based on Shannon diversity and Simpson diversity. The ratio of observed diversity to maximum diversity is taken as the measure of evenness (E). Two different evenness measures were calculated, the Shannon Evenness (E1) and Sheldon Evenness (E2). Evenness indices of bird community recorded in different months are given in Table 8.3. Highest evenness was obtained during July (0.80) and June (0.72).

Table 8.3: Evenness indices of bird community in the Kole wetlands during different months

Months	Shannon Evenness (E1)	Sheldon Evenness (E2)
January	0.65	0.21
February	0.53	0.13
March	0.63	0.18
April	0.60	0.17
May	0.64	0.24
June	0.72	0.35
July	0.80	0.49
August	0.65	0.26
September	0.57	0.14
October	0.46	0.09
November	0.55	0.13
December	0.63	0.17

Density

As observed in the species richness and abundance of birds, density of birds also varied in different months. Highest density of 29,158 birds/ha was recorded in December followed by November (24,373 birds/ha). Lowest density of birds was observed in July (272). Density of birds in different months and standard error is presented in Fig. 8.6.

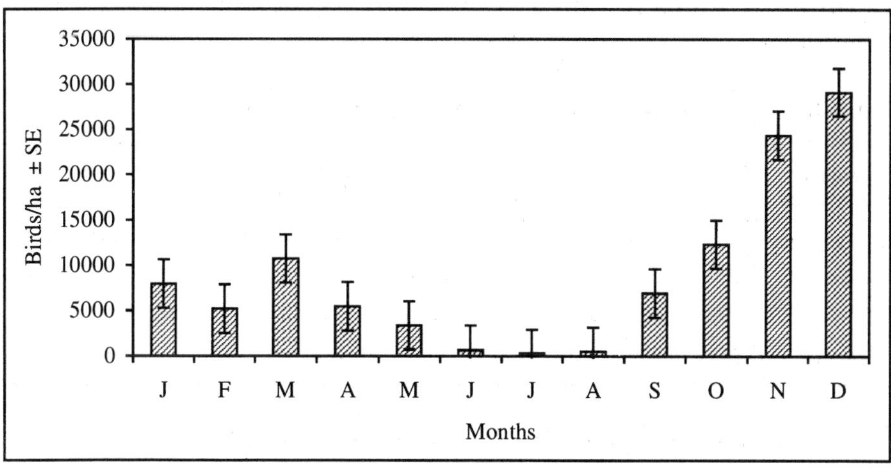

Fig. 8.6: Density (birds/ha) of birds in different months in the Kole wetlands (1998-2001)

Species richness in the intensive study areas

Species richness of birds varied in different months in the intensive study areas. Highest number of species was recorded at Parappur during December (79) and lowest was at Chettupuzha during June (10). The number of species increased during the winter months and decreased during monsoon (Table 8.4).

Table 8.4: Species richness of birds in the intensive study areas

Intensive study areas	Months											
	J	F	M	A	M	J	J	A	S	O	N	D
Chettupuzha	30	28	20	25	17	10	17	21	22	28	39	30
Kanjany	58	58	63	51	28	17	22	27	36	45	65	76
Enamavu	32	21	37	25	29	14	21	20	46	36	30	27
Parappur	42	44	69	53	33	38	32	32	53	63	56	79

Species Richness Index R1 was highest at Parappur (8.25) in December followed by Kanjany (6.69). Overall Species Richness R1 and R2 were highest at Parappur and Kanjany.

Abundance of birds in the intensive study areas

Total number of birds in the intensive study areas varied from 140 to 73,604 individuals in a month. Highest number of birds was recorded at Kanjany during December (73,604) and lowest was at Chettupuzha during July (140) (Table 8.5).

Table 8.5: Abundance of birds in the intensive study areas

Intensive study areas	Months											
	J	F	M	A	M	J	J	A	S	O	N	D
Chettupuzha	1456	1680	982	753	460	208	140	301	733	544	9871	1230
Kanjany	25514	21501	20110	11012	2466	353	346	903	5881	11603	70683	73604
Enamavu	1994	707	4144	1531	4353	1453	156	408	20351	6864	4715	3649
Parappur	2912	7780	18799	9111	1274	739	486	425	1509	31146	13452	14380

Diversity indices (H') in the intensive study areas

The highest diversity Index (H′) was observed in June (2.90) at Parappur and the lowest diversity Index in February (1.53) at Parappur

(Table 8.6). Similarly, the lowest and the highest diversity Index were recorded at Parappur.

Table 8.6: Diversity indices (H') of birds in the intensive study areas

Intensive study areas	Months											
	J	F	M	A	M	J	J	A	S	O	N	D
Chettupuzha	2.33	2.20	2.02	2.37	2.03	1.84	2.08	2.14	2.00	2.61	2.15	2.43
Kanjany	2.81	2.51	2.37	2.02	2.36	2.39	2.64	1.90	2.40	1.87	2.28	2.84
Enamavu	2.31	2.18	2.41	1.99	2.04	1.62	2.51	2.18	2.22	2.16	2.04	1.98
Parappur	2.37	1.53	2.27	2.31	2.20	2.90	2.69	2.88	2.33	1.78	2.52	2.63

Density of birds in the intensive study areas

The highest density of birds was recorded in December (53,994 birds/ha) at Kanjany and the lowest was in July (103 birds/ha) at Chettupuzha (Table 8.7).

Table 8.7: Density of birds in the intensive study areas (Number of birds/ha)

Intensive study areas	Months											
	J	F	M	A	M	J	J	A	S	O	N	D
Chettupuzha	1068	1232	720	552	337	153	103	221	538	399	741	902
Kanjany	16516	15772	14752	8078	1809	259	254	662	4324	8512	51851	53994
Enamavu	1463	519	3040	1123	3193	1066	114	299	14929	5035	3459	2677
Parappur	2136	5707	13790	7784	935	542	357	312	1107	22848	9848	10549

Similarity Indices between the intensive study areas

Two similarity indices namely Jaccard Index and Sorenson Index were calculated. An alternative approach to measure the similarity of different sites is using similarity indices, using the proportional similarity measures. Similarity indices between the intensive study areas were computed using qualitative data (Table 8.8). Both indices showed highest similarity between Enamavu and Kanjany (0.62 and 0.76). Similarity between Chettupuzha, Kanjany and Parappur was less than fifty per cent.

Table 8.8: Jaccard Index and Sorenson Index values (in parenthesis) for the intensive study areas

Intensive study areas	Chettupuzha	Enamavu	Kanjany	Parappur
Chettupuzha	0 (0)	0.56 (0.72)	0.46 (0.62)	0.43 (0.60)
Enamavu		0 (0)	0.62 (0.76)	0.48 (0.52)
Kanjany			0 (0)	0.51 (0.67)
Parappur				0 (0)

Distribution model

Another way of describing diversity in a community is through species-abundance or distribution model introduced by Fisher *et al.* (1943). A species-abundance model utilises all information gathered in a community and is the most complete mathematical description of the data (Magurran, 1988). The data analysis showed that the truncated lognormal model is fitting to the bird community at the Kole wetlands.

The distribution model indicates the absence of a single dominant species or group of species and the presence of long series of very rare species at Kole wetlands. The species, which is represented by less than 2 individuals, can be called as rare. The observed and expected number of species was compared using the χ^2 goodness of fit test. The test showed that there is no significant difference between the observed and expected distribution (χ^2 = 18.31; P = 0.08). Fig. 8.7 and Table 8.9 indicate that the bird community is following the truncated lognormal distribution pattern.

Overall bird community parameters are presented in Table 8.10. Diversity Index (H′) was 3.11 and (λ) 0.08. Species Richness Index R1 was 13.96 and R2 was 0.28. Similarly, high values were obtained for Hill's number N1 and N2. Hill's number N1 was 22.38 and Hill's numbers N2 was 12.36. Evenness index (E1) was 0.60 and (E2) 0.12.

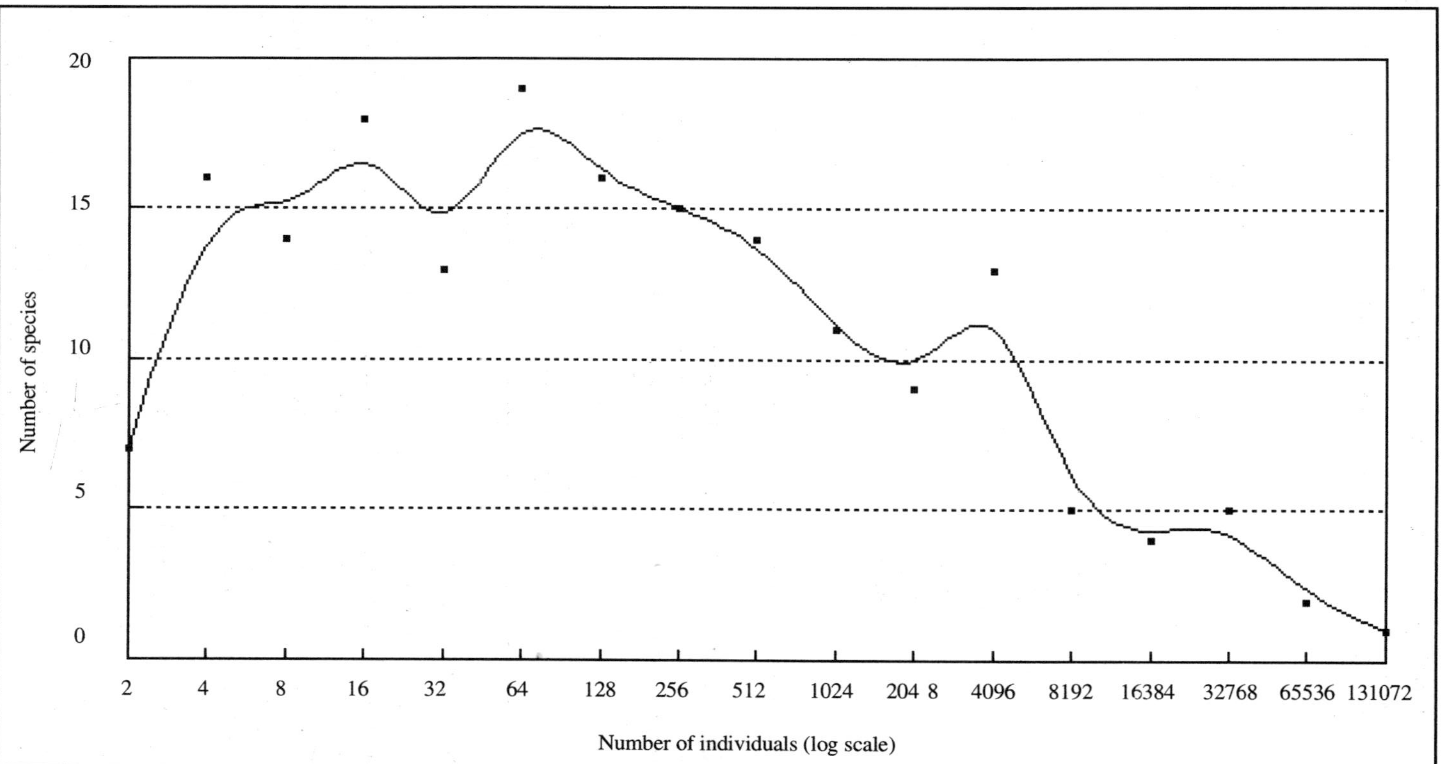

Fig. 8.7: Species-abundance distribution of birds at Kole wetlands

Table 8.9: Truncated lognormal distribution at Kole wetlands (χ^2 test)

Class	Upper boundary	Observed	Expected	χ_2
Behind veil line	0.5	--	7.21	--
1	2.5	7	14.92	4.20
2	4.5	16	9.15	5.13
3	8.5	14	12.51	0.18
4	16.5	18	15.09	0.56
5	32.5	13	18.02	1.40
6	64.5	19	20.34	0.09
7	128.5	16	20.01	0.80
8	256.5	15	20.08	1.29
9	512.5	14	17.49	0.70
10	1024.5	11	15.59	1.35
11	2048.5	9	12.04	0.77
12	4096.5	13	9.52	1.27
13	8192.5	5	6.74	0.45
14	∝	12	10.82	0.13
	Total	**182**	**209.53**	**18.31**

Table 8.10: Bird community parameters in the Kole wetlands

Species Richness Index		Shannon Index	Simpson's Index	Hill's Number		Evenness Index	
R1	R2	H'	λ	N1	N2	E1	E2
13.96	0.28	3.11	0.08	22.38	12.36	0.60	0.12

Wetland bird community

Birds of Kole wetlands fall into six Orders. Highest number of birds was recorded from the Order Charadriiformes (1,52,727) followed by Ciconiiformes (1,33,491) and Pelecaniformes (42,532) (Fig. 8.8). Sixteen Families of wetland species were recorded during the study, out of this maximum number of species was from the family Ardeidae (1,27,684) followed by Laridae (96,940) (Fig. 8.9).

Dominance in wetland bird species

Out of 82 wetland bird species observed in the Kole wetland, Whiskered Tern (23 per cent) was highest in dominance followed by Little Egret (13 per cent) and Little Cormorant (11 per cent) (Table 8.11). Sixty-eight species were represented in less than 1 per cent. The Chestnut Bittern, Eurasian Woodcock, Slaty-legged Crake, Indian Moorhen, Painted Stork, Eurasian Spoonbill, Spot-billed Pelican, European White Stork, Indian Shag, Black Stork and Little Green Heron were of low dominance. Among the migratory waders, Wood Sandpiper, Little Stint and Little Ringed Plover were the most abundant species.

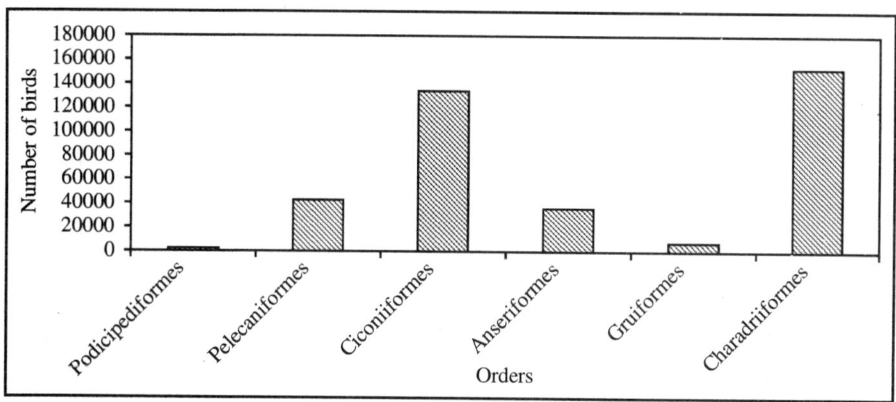

Fig. 8.8: Order wise abundance of wetland bird species (n = 36 months)

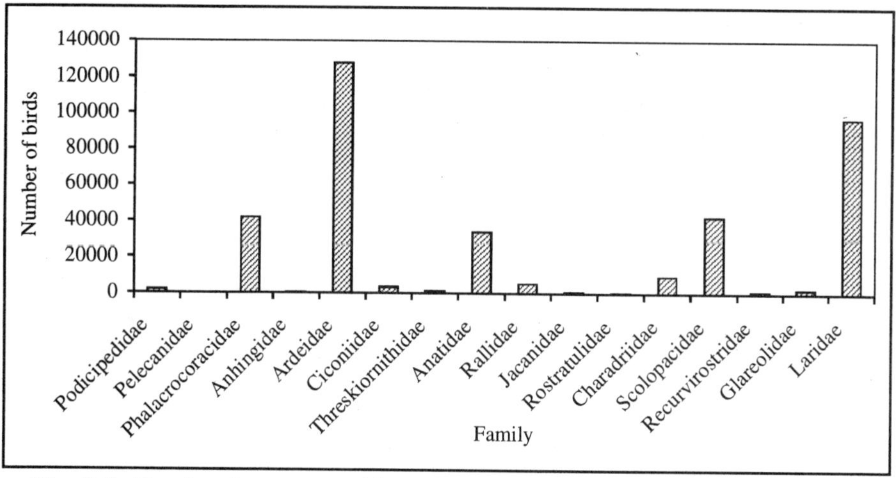

Fig. 8.9: Family wise composition of wetland bird species (n = 36 months)

Herons and Egrets : Out of the 12 species of egrets and herons, Little Egret was abundant (37 per cent) followed by Cattle Egret (24 per cent) (Table 8.12). Species like Black-crowned Night-Heron, Purple Heron, Western Reef Egret, Yellow Bittern, Black Bittern, Chestnut Bittern and Little Green Heron were recorded only less than 1 per cent.

Table 8.11: Abundance and dominance of wetland species in the Kole wetlands (+ = dominance less than 0.01)

Sl. No.	Common name	Abundance	Sighting frequency	Dominance Index
1.	Whiskered Tern	85540	282	23.0
2.	Little Egret	47938	261	12.8
3.	Little Cormorant	39691	360	10.7
4.	Cattle Egret	29703	215	7.9
5.	Median Egret	24785	156	6.7
6.	Garganey	21513	10	5.8
7.	Indian Pond-Heron	21214	387	5.7
8.	Wood Sandpiper	13871	151	3.7
9.	Little Stint	12846	39	3.5
10.	Northern Pintail	9204	6	2.5
11.	Black-headed Gull	6889	4	1.9
12.	Little Ringed Plover	6536	43	1.8
13.	Purple Moorhen	4604	45	1.2
14.	Brown-headed Gull	4502	4	1.2
15.	Common Sandpiper	3274	103	0.9
16.	Curlew Sandpiper	3206	25	0.9
17.	Asian Openbill-Stork	3174	21	0.9
18.	Great Cormorant	2700	1	0.7
19.	Large Egret	2635	25	0.7
20.	Eurasian Curlew	2404	25	0.7
21.	Small Pratincole	2253	7	0.6
22.	Little Grebe	2167	103	0.6
23.	Ruff	2103	28	0.6
24.	Lesser Whistling-Duck	1708	70	0.5
25.	Lesser Sand Plover	1464	15	0.4
26.	Common Teal	1319	9	0.4
27.	Marsh Sandpiper	1305	67	0.4
28.	Grey Heron	1276	64	0.3
29.	Oriental White Ibis	1230	13	0.3
30.	Red-wattled Lapwing	1186	170	0.3
31.	Green Sandpiper	935	17	0.3

Contd...

Table Contd...

32.	Temminck's Stint	820	11	0.2
33.	Common Coot	702	25	0.2
34.	Black-winged Stilt	665	19	0.2
35.	Common Green Shank	612	34	0.2
36.	Gadwall	579	1	0.2
37.	Black-crowned Night- Heron	500	2	0.1
38.	Broad-billed Sandpiper	441	8	0.1
39.	Purple Heron	434	78	0.1
40.	Pacific Golden-Plover	346	4	0.1
41.	Spot-billed Duck	335	11	0.1
42.	Cotton Teal	317	25	0.1
43.	Caspian Tern	316	2	0.1
44.	Western Reef Egret	294	23	0.1
45.	Black-tailed Godwit	242	7	0.1
46.	Sanderling	222	2	0.1
47.	Dunlin	216	3	0.1
48.	Pheasant-tailed Jacana	200	12	0.1
49.	White-breasted Waterhen	200	67	0.1
50.	Bronze-winged Jacana	121	16	0.03
51.	Darter	117	27	0.03
52.	Ferruginous Pochard	82	1	0.02
53.	Pied Avocet	70	3	0.02
54.	Kentish Plover	62	4	0.02
55.	Bar-tailed Godwit	60	3	0.02
56.	Northern Shoveller	60	3	0.02
57.	White-necked Stork	58	10	0.02
58.	Greater Painted-Snipe	54	7	0.01
59.	Ruddy Turnstone	53	2	0.01
60.	Common Redshank	50	7	0.01
61.	Ruddy-breasted Crake	47	11	0.01
62.	Yellow-legged Gull	45	2	0.01
63.	Pintail Snipe	41	4	0.01
64.	Yellow Bittern	41	15	0.01
65.	Black Ibis	38	13	0.01

Contd...

Table Contd...

66.	European White Stork	36	1	0.01
67.	Common Snipe	34	9	0.01
68.	Whimbrel	34	2	0.01
69.	Black Bittern	33	21	0.01
70.	Watercock	32	9	0.01
71.	Great Knot	24	4	0.01
72.	Terek Sandpiper	22	4	0.01
73.	Chestnut Bittern	16	9	+
74.	Eurasian Woodcock	15	6	+
75.	Slaty-legged Crake	9	2	+
76.	Common Moorhen	9	3	+
77.	Painted Stork	5	3	+
78.	Eurasian Spoonbill	4	1	+
79.	Spot-billed Pelican	4	2	+
80.	Indian Shag	2	1	+
81.	Black Stork	1	1	+
82.	Little Green Heron	1	1	+
	Total	**3,71,896**		**100**

Table 8.12: Dominance of Herons and Egrets recorded from the Kole wetlands (n=36) (+ = dominance less than 1)

S. No.	Species	Abundance	Sighting frequency	Dominance Index
1.	Little Egret	46938	261	37.4
2.	Cattle Egret	29703	215	23.6
3.	Median Egret	24785	156	19.7
4.	Indian Pond -Heron	21214	387	16.9
5.	Grey Heron	1276	64	1.02
6.	Black-crowned Night-Heron	943	2	+
7.	Purple Heron	434	78	+
8.	Western Reef -Egret	294	23	+
9.	Yellow Bittern	41	15	+
10.	Black Bittern	39	21	+
11.	Chestnut Bittern	16	9	+
12.	Little Green Heron	1	1	+
	Total	**1,25,684**		**100**

Storks and Ibis : Eight species of Storks and Ibis were recorded from the Kole wetlands. Among these, highest dominance index was for Asian Openbill Stork (70 per cent) followed by Oriental White Ibis (27 per cent) (Table 8.13).

Table 8.13: Dominance of Storks and Ibis recorded from the Kole wetlands (n=36)

S. No.	Species	Abundance	Sighting frequency	Dominance index
1.	Asian Openbill - Stork	3174	21	69.8
2.	Oriental White Ibis	1230	13	27.1
3.	White-necked Stork	58	10	1.3
4.	Black Ibis	38	13	0.8
5.	European White Stork	36	1	0.7
6.	Painted Stork	5	3	0.1
7.	Eurasian Spoonbill	4	1	0.1
8.	Black Stork	1	1	0.02
	Total	**4546**		**100**

Ducks : Out of the nine ducks, Garganey (59 per cent) and Northern Pintail (25 per cent) were abundant, whereas other species were only few in numbers and Northern Shoveller was the lowest in abundance. Four species showed less than 1 per cent dominance. Abundance of different species of ducks recorded from the Kole wetlands is given in (Table 8.14).

Table 8.14: Dominance index of different species of ducks recorded from the Kole wetlands (n=36)

S. No.	Species	Abundance	Sighting frequency	Dominance Index
1.	Garganey	21513	10	59.2
2.	Northern Pintail	9204	6	25.3
3.	Lesser Whistling-Duck	3044	70	8.4
4.	Common Teal	1319	9	3.6
5.	Gadwall	473	1	1.3
8.	Spot-billed Duck	335	11	0.9
7.	Cotton Teal	317	25	0.9
6.	Ferruginous Pochard	82	1	0.2
9.	Northern Shoveller	60	3	0.2
	Total	**36,347**		**100**

Egrets and allied species : Egrets form one of the foremost groups of birds recorded from the Kole wetlands. Large flocks of egrets in white plumage provided a splendid appearance to the Kole wetlands. However, not all the species of egrets were represented in equal numbers. More than 35,000 Little Egrets per/ha were observed in the Kole wetlands, which indicated the abundance of this species in the region. Density of egrets and other related species is given in (Table 8.15).

Table 8.15: Density of Egrets and allied species recorded from the Kole wetlands

S. No.	Common name	Number of birds/ha
1.	Little Egret	35023
2.	Little Cormorant	28642
3.	Cattle Egret	21612
4.	Median Egret	18165
5.	Indian Pond-Heron	15124
6.	Purple Moorhen	3192
7.	Asian Openbill-Stork	2328
8.	Large Egret	1911
9.	Grey Heron	916
10.	Oriental White Ibis	902
11.	Black-crowned Night- Heron	692
12.	Purple Heron	282
13.	Western Reef-Egret	216
14.	White-breasted Waterhen	139
15.	Darter	84
16.	White-necked Stork	43
17.	Ruddy-breasted Crake	32
18.	Yellow Bittern	30
19.	Watercock	15
20.	Chestnut Bittern	12

Feeding guilds

Feeding guild wise analysis of birds showed that species richness was highest in the insectivores followed by aquatic feeders throughout the study period. Abundance of birds was highest in the aquatic feeders and omnivores (Fig. 8.10).

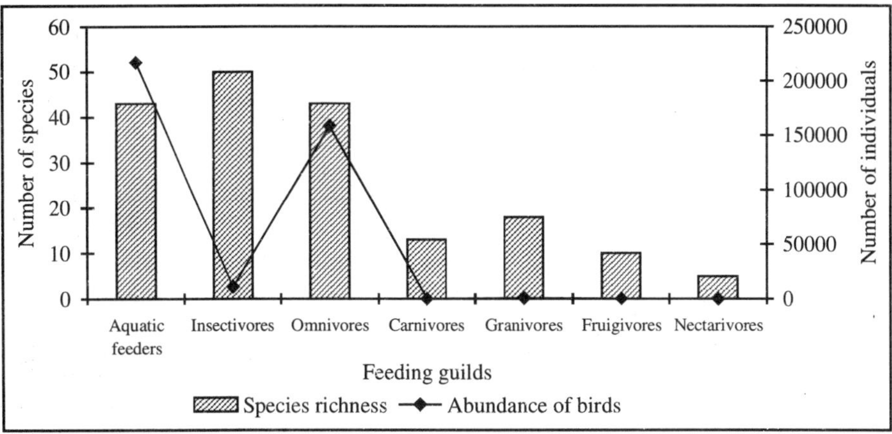

Fig. 8.10: Feeding guild wise comparison of bird species richness and abundance of birds

During this study, birds were observed 5,382 times in which 4,32,663 birds were counted. Species richness and abundance of birds showed high values in the Kole wetlands, which is comparable to other wetlands in Kerala (Kurup, 1996; Jayson and Easa, 2000) and in India (Sampath, 1989; Nagarajan and Thiyagesan, 1996). The highest number of birds was recorded during November, which showed the influx of birds into the region due to the trans-continental migration. Population was low during June to July, when the migratory species were absent and the few resident species moved away to avoid the heavy rain. As the whole wetland lay inundated during this period, availability of food was also low. Only diving species like Little Cormorant, Darter and Little Grebe preferred the area during the months of southwest monsoon. Occurrence of 97 species of birds in a month is commendable, which showed the importance of the area for the migratory species. The increase in wetland species from September to March implies the presence of their preferred microhabitat and higher production of benthic and macro fauna. As reported earlier from the Western Ghats, highest number of birds was recorded during the winter and there was a reduction in population size during the monsoon (Daniels, 1989).

Species richness in an area is dependent on the availability of food, climate, evolutionary history and predation pressure. Diversity indices are dependent on two factors, species richness and evenness. It is directly correlated with the stability of the ecosystem. It will be higher in the biologically controlled systems and low in disturbed ecosystem.

In the Kole wetlands, diversity indices were higher. As the evenness measures also showed high values, it could be concluded that species of individuals uniformly presented and this indicated the conservation value of the wetlands.

All the egret species showed uniform pattern of population fluctuations. Herons and Egrets are recognized as important biological indicators of environmental change in wetlands. Reasons for this recognition include, their position as top carnivores, which can signal changes occurring at lower trophic levels; communal nesting which facilitates sampling of reproductive output; use of human altered landscapes; sensitivity to disturbance and pollution; and dependence on specific hydrologic and hydrographic functions with associated responsiveness to changes in fundamental physical ecosystem characteristics.

Among the four intensive study sites, species richness and diversity was highest at Parappur and abundance and density of birds was highest at Kanjany. The geographic position of Kanjany is in the middle of the Kole wetlands. Many models are available for describing the species-abundance distribution. Preston (1948) introduced the lognormal distribution to explain the species-abundance data. Usually in ecological work, distribution of species is always truncated at the left side (Preston, 1962). Geometric series patterns are usually found in species poor or harsh environments. Log series patterns are usually observed where one or a few factors dominate the ecology of a community. Lognormal distribution is found in most biological populations. The Broken-stick model distribution shows the maximum equitable distribution of available resources. Species-abundance distribution at Kole wetlands followed the truncated lognormal model, which indicated the presence of natural bird community in the area. The Kole Wetlands is one of the important staging and wintering areas for migratory waterfowl in the Central Asian Indian Flyway. This region supported waders similar to the known habitats such as Chilika Lake, Pulicate Lake and Great Vedaranyam Swamp.

A number of hypotheses have been forwarded to explain the characteristic diversity profiles of different habitats. Habitat heterogeneity, in addition to the area, is an important determinant of species richness. Shannon Index obtained for the area is comparable with other wetlands. Even though the total number of birds and species richness reduced during the southwest monsoon, it was not reflected in the diversity indices. All the study years showed high diversity index values. Evenness indices indicated high values during June to July, when the abundance of birds was lowest. High similarity was observed between Kanjany and Enamavu. This is because these two intensive study sites are continuous stretch of the Kole wetlands. High abundance of insectivores in this wetland showed the importance of wetlands to the insectivorous bird species.

According to Francklin (1989) long term studies are needed in ecology to understand the population dynamics. Despite a number of such studies on birds, controversy remains as to the nature and role of density dependent and density independent factors in regulating or controlling the population size (Krebs, 1991; Owen and Black, 1991). Density dependence is manifested in a reduced capacity for further increase in numbers as population size grows, and is the mechanism that stabilizes the population, that might otherwise increase or decrease without bounds (Newton, 1998). In recent years, many animal species throughout the world are threatened with extinction or are becoming increasingly endangered due to anthropogenic and other factors. Because of the natural and anthropogenic factors in the habitat, forecasting the effects of habitat degradation or impending global climate change on the dynamics of animal populations, community and biodiversity is of worldwide concern. Such studies are important because naturally occurring levels of biodiversity are critical to preserve, not only because of the potential economic or medical benefits yet to be uncovered, but also because, maintaining biodiversity is synonymous with maintaining an intact ecosystem within which we too live (Soule and Wilcox, 1980; Wilson, 1985).

❑❑❑

Chapter 9

Seasonal Changes of the Bird Community

Population studies have been traditionally used to monitor large scale, long term changes in avian population and to assess both habitat quality and the responses of birds to both management practices and natural and human caused environmental changes (Wiens, 1989). Species composition, abundance and behaviour of birds are known to vary seasonally (Morrison *et al.* 1980). Some species are permanent residents in an area, others occupy an area only during winter or summer months. In this chapter, seasonal variations of wetland bird community in the Kole wetlands are described in terms of species richness, abundance, diversity and density.

METHODS

Seasonal pattern

The study area has three distinct seasons *viz.* dry season (December to April) wet-I, the period of southwest monsoon (May to August) and wet-II, the period of northeast monsoon (September to November). Based on the arrival of migratory birds, the months are divided into migratory season (September to March) and non-migratory season (April to August). Occurrence of birds in each month obtained through the daily census data was used for the seasonality analysis. Monthly changes and seasonal changes of the community were analysed. Species richness, abundance, diversity indices of each season and also in migratory and non-migratory season were calculated and presented. Population fluctuations of 36 wader species in the four intensive study areas during three migratory seasons were estimated. Apart from these, detailed information on the population fluctuation of selected bird species was presented.

Correlation coefficient was used to find out the correlation between the bird species richness, abundance and diversity with environmental factors like rainfall and water depth. Paddy growth was measured using centimeter scale in different stages and correlated with the abundance of birds.

Seasonal changes of the bird community

Species of birds recorded in each month during the period of study is given in Table 9.1. Out of the one hundred and eighty two species, fourteen species were recorded in all months, four species in 11 months, eleven species in 10 months, 9 species in nine and eight months, 13 in seven months, 14 in six months, 16 in five months, 18 in four months, 29 in three months, 25 species in two months and 20 species only in one month.

Table 9.1: Occurrence of birds in different months in the Kole wetlands (n=36)

S. No.	Common name	J	F	M	A	M	J	J	A	S	O	N	D
1.	Little Grebe	√	√	√	√	√	√	√	√	√	√		√
2.	Spot-billed Pelican											√	√
3.	Masked Booby											√	
4.	Little Cormorant	√	√	√	√	√	√	√	√	√	√	√	√
5.	Great Cormorant	√	√	√	√			√	√	√	√	√	√
6.	Indian Shag								√				
7.	Darter	√	√	√	√				√	√	√	√	√
8.	Lesser Frigatebird						√						
9.	Grey Heron	√	√	√	√			√		√	√	√	√
10.	Purple Heron	√	√	√	√	√	√	√	√	√	√	√	√
11.	Little Green Heron									√			

Contd...

Table Contd...

12.	Indian Pond-Heron	√	√	√	√	√	√	√	√	√	√	√	√
13.	Cattle Egret	√	√	√	√	√	√		√		√	√	√
14.	Large Egret		√		√	√				√			
15.	Median Egret	√	√	√	√	√			√	√	√	√	√
16.	Little Egret	√	√	√	√	√			√	√	√	√	√
17.	Western Reef-Egret	√		√	√						√	√	√
18.	Black-crowned Night-Heron	√	√										
19.	Chestnut Bittern			√	√	√				√			√
20.	Yellow Bittern	√	√	√	√	√	√		√	√		√	
21.	Black Bittern												
22.	Painted Stork	√	√	√	√	√	√	√	√		√	√	
23.	Asian Openbill-	√											√
	Stork	√	√	√	√	√						√	√
24.	White-necked Stork			√	√	√						√	√
25.	European White	√										√	√
	Stork	√											
26.	Black Stork												
27.	Oriental White Ibis			√	√	√				√	√	√	√
28.	Black Ibis									√	√	√	
29.	Eurasian Spoonbill												√
30.	Lesser Whistling-Duck	√		√	√	√	√	√	√	√	√		√
31.	Common Teal			√						√	√		√
32.	Northern Pintail	√		√								√	√
33.	Spot-billed Duck		√	√	√								√
34.	Gadwall				√							√	√
35.	Garganey		√	√	√							√	
36.	Northern Shoveller		√										√
37.	Ferruginous Pochard											√	
38.	Cotton Teal			√	√	√	√	√	√		√		
39.	Black-shouldered Kite	√	√	√	√	√						√	√
40.	Black Kite	√							√	√	√	√	√
41.	Brahminy Kite	√		√	√	√		√		√	√	√	
42.	Shikra							√					√
43.	Pallid Harrier		√	√							√		
44.	Pied Harrier	√	√	√								√	√
45.	Western Marsh-Harrier	√	√	√	√	√				√	√	√	√

Contd...

Table Contd...

46.	Eurasian Sparrowhawk			√								√	
47.	Oriental Honey-Buzzard			√		√							
48.	Osprey											√	√
49.	Grey Francolin			√	√								
50.	Red Spurfowl	√								√	√		
51.	Indian Peafowl									√	√		√
52.	Ruddy-breasted Crake	√	√	√	√	√	√						
53.	Slaty-legged Crake	√	√	√		√							
54.	White-breasted Waterhen	√	√	√	√	√	√	√			√	√	√
55.	Watercock			√	√	√	√					√	√
56.	Common Moorhen			√				√					
57.	Purple Moorhen	√	√	√	√	√	√	√	√	√	√	√	√
58.	Common Coot					√	√			√			
59.	Pheasant-Tailed Jacana			√	√	√			√		√	√	
60.	Bronze-winged Jacana	√		√	√	√		√		√		√	√
61.	Greater Painted-Snipe	√	√		√	√							
62.	Black-winged Stilt	√	√	√	√						√	√	√
63.	Pied Avocet											√	√
64.	Small Pratincole	√								√	√	√	√
65.	Red-wattled Lapwing	√	√	√	√	√	√	√	√	√	√	√	√
66.	Pacific Golden-Plover	√	√									√	√
67.	Lesser Sand Plover									√	√	√	√
68.	Little Ringed Plover	√	√	√	√					√	√	√	√
69.	Kentish Plover									√	√		
70.	Black-tailed Godwit									√	√	√	
71.	Bar-tailed Godwit		√							√			√
72.	Whimbrel									√	√		√
73.	Eurasian Curlew	√	√							√		√	√
74.	Common Redshank	√	√							√	√	√	√
75.	Marsh Sandpiper	√	√									√	√
76.	Common Greenshank	√	√	√	√						√	√	√

Contd...

Table Contd...

77.	Green Sandpiper	√	√	√	√					√	√	√	√
78.	Wood Sandpiper	√	√	√	√	√				√	√	√	√
79.	Terek Sandpiper											√	√
80.	Common Sandpiper	√	√	√	√					√	√		
81.	Ruddy Turnstone									√			
82.	Great Knot										√	√	
83.	Dunlin									√	√		
84.	Curlew Sandpiper		√							√	√	√	√
85.	Broad-billed Sandpiper									√	√	√	√
86.	Common Snipe	√		√	√							√	√
87.	Pintail Snipe	√		√									√
88.	Sanderling												√
89.	Little Stint	√	√							√	√	√	√
90.	Temminck's Stint	√								√		√	√
91.	Ruff	√	√							√	√	√	
92.	Eurasian Woodcock												√
93.	Yellow-legged Gull	√											√
94.	Brown-headed Gull	√										√	√
95.	Black-headed Gull	√										√	√
96.	Whiskered Tern	√	√	√	√	√	√	√	√	√	√	√	√
97.	Caspian Tern									√			
98.	Blue Rock Pigeon	√		√	√					√			√
99.	Spotted Dove	√	√	√	√			√		√	√	√	√
100.	Eurasian Collared Dove				√						√		√
101.	Rose-ringed Parakeet		√		√								√
102.	Plum-headed Parakeet				√					√		√	√
103.	Pied Crested Cuckoo			√	√				√				
104.	Brainfever Bird					√					√		
105.	Indian Cuckoo		√	√	√								√
106.	Banded Bay Cuckoo												√
107.	Asian Koel	√	√	√	√						√	√	√
108.	Greater Coucal	√	√	√	√	√	√	√	√		√	√	√
109.	Barn Owl	√	√										
110.	Spotted Owlet					√							√
111.	Mottled Wood-Owl	√	√										√
112.	Alpine Swift			√									
113.	House Swift										√	√	√
114.	Asian Palm Swift			√					√				

Contd...

Table Contd...

115.	Lesser Pied Kingfisher	√	√	√	√	√	√	√	√	√	√	√	√
116.	Small Blue Kingfisher	√	√	√	√	√	√	√	√	√	√	√	√
117.	Storkbilled Kingfisher	√	√	√	√	√	√	√	√	√	√	√	√
118.	White-breasted Kingfisher	√	√	√	√	√	√	√	√	√	√	√	√
119.	Black-capped Kingfisher	√		√	√							√	√
120.	Blue-tailed Bee-Eater	√	√	√		√	√				√	√	√
121.	Small Bee-Eater		√	√			√	√	√	√	√	√	√
122.	Indian Roller	√		√	√	√	√	√		√	√	√	√
123.	Common Hoopoe	√										√	√
124.	White-cheeked Barbet		√		√	√	√	√		√			√
125.	Lesser Golden-backed Woodpecker	√	√	√		√	√			√			√
126.	Common Swallow	√	√	√								√	√
127.	House Swallow	√		√							√	√	√
128.	Red-rumped Swallow	√	√	√	√							√	√
129.	Brown Shrike	√	√	√									
130.	Eurasian Golden Oriole	√	√	√	√	√	√			√			√
131.	Black-headed Oriole	√	√	√			√			√	√	√	
132.	Black Drongo	√	√	√	√	√	√	√	√		√	√	√
133.	Ashy Drongo			√	√			√		√	√		√
134.	White-bellied Drongo	√	√	√							√	√	√
135.	Ashy Wood Swallow	√	√	√		√	√	√	√	√	√		
136.	Common Myna	√	√	√	√	√	√	√	√	√	√	√	√
137.	Jungle Myna	√	√	√		√	√	√	√	√	√	√	
138.	Grey-headed Starling						√	√	√	√			
139.	Indian Tree Pie	√	√	√	√	√	√	√	√	√	√	√	√
140.	House Crow	√	√	√	√	√	√	√	√	√	√	√	√
141.	Jungle Crow	√		√									
142.	Common Iora	√											√
143.	Gold-fronted Chloropsis			√									
144.	Jerdon's Chloropsis												√

Contd...

Table Contd...

145. Red-whiskered Bulbul	√	√		√								
146. Red-vented Bulbul		√	√	√	√	√	√	√	√	√	√	
147. White-headed Babbler	√	√	√	√		√			√		√	√
148. Common Babbler					√	√			√	√	√	√
149. Jungle Babbler		√	√	√								
150. Indian Rufous Babbler	√			√								√
151. Asian Paradise-Flycatcher												√
152. Streaked Fantail-Warbler		√						√	√		√	
153. Franklin's Prinia				√								
154. Plain Prinia		√	√	√				√	√	√		√
155. Ashy Prinia	√	√	√	√	√	√	√	√	√	√	√	√
156. Common Tailorbird	√	√	√	√	√			√	√	√	√	√
157. Indian Great Reed-Warbler	√	√	√	√	√	√	√				√	√
158. Blyth's Reed Warbler											√	
159. Oriental Robin										√		√
160. Pied Bushchat	√									√		√
161. Desert Wheatear	√		√	√					√	√		
162. Indian Robin					√	√			√	√		
163. Paddy Field Pipit			√	√						√	√	√
164. Eurasian Tree Pipit										√	√	√
165. Yellow Wagtail	√	√	√	√						√		
166. Citrine Wagtail	√										√	√
167. Grey Wagtail	√										√	√
168. Large Pied Wagtail			√	√					√		√	√
169. Tickell's Flowerpecker									√		√	
170. Thick-billed Flowerpecker			√	√								
171. Purple-rumped Sunbird			√						√			√
172. Purple Sunbird	√								√	√		√
173. Loten's Sunbird			√									
174. Black breasted Weaver	√											
175. Baya Weaver	√	√	√	√	√				√		√	√
176. Streaked Weaver	√	√	√	√								√

Contd...

Table Contd...

177. Red Munia	√	√		√				√	√	√	√	√
178. White-rumped Munia		√	√						√			
179. White-throated Silverbill	√	√	√		√						√	√
180. Black-throated Munia		√	√	√				√	√	√	√	
181. Spotted Munia		√	√	√	√	√			√			
182. Black-headed Munia	√	√	√	√	√	√		√	√	√	√	√

√ = Present

Species richness and abundance : Species richness and abundance of birds varied in different seasons. Species richness and abundance were highest during the dry season of 2000 and lowest in the wet-I season of 1999 (Fig. 9.1). The wet-I season always recorded lowest species richness and abundance of birds.

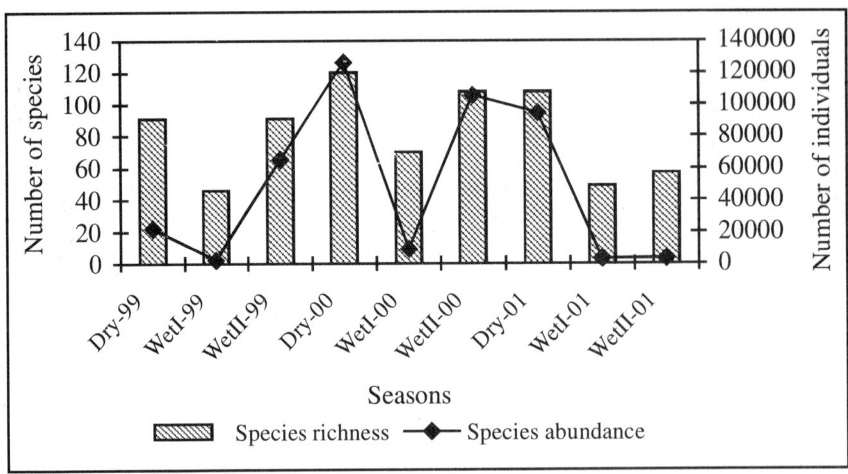

Fig. 9.1: Species richness and abundance of birds in different seasons

Diversity Index : Diversity Index (H′) was highest in the dry season of 2000 (3.19) followed by the dry season of 2001 (2.89) (Fig. 9.2). The lowest diversity index (H′) was observed in wet-II season (2.45).

Migratory and non-migratory seasons

Species richness was highest in the third year migratory season (2000-2001) followed by second year (1999-2000). Highest abundance was in the third year migratory season (2000-2001) and lowest in the first year (1998-1999) (Fig. 9.3 and 9.4). Species richness and abundance was high during the migratory season compared to the non-migratory season.

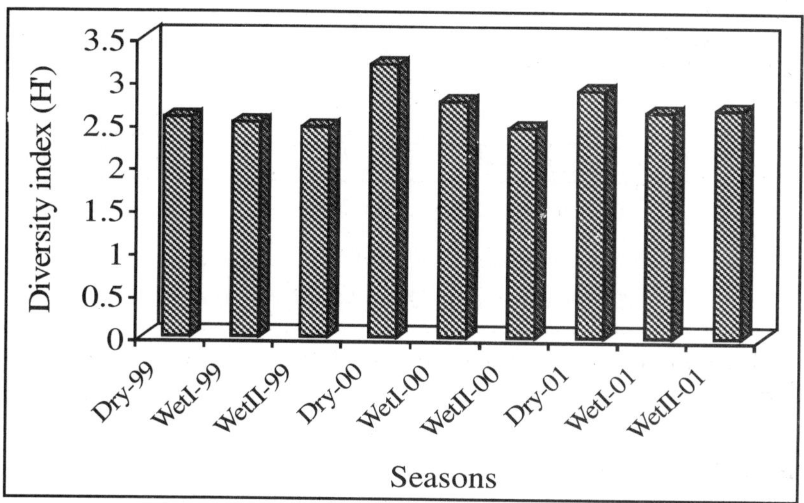

Fig. 9.2: Diversity Index (H') of birds in different seasons

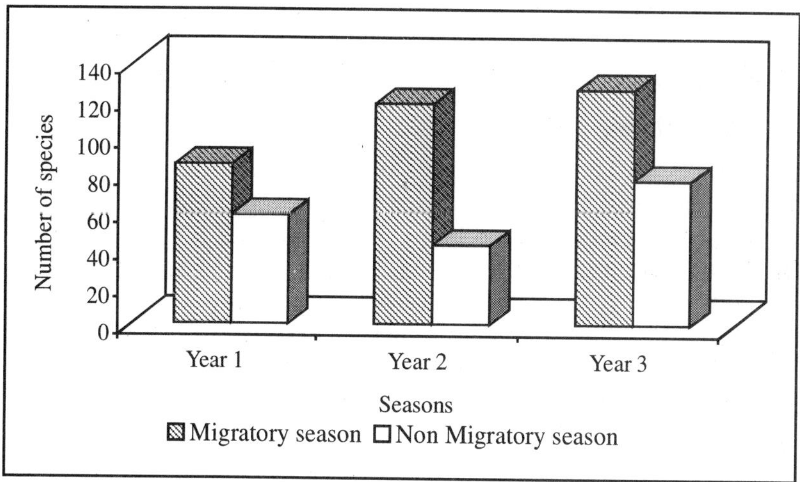

Fig. 9.3: Species richness in migratory and non-migratory seasons (1998-2001)

Highest diversity index (H′) was in the second year migratory season (1999-2000) and lowest in the second year non-migratory season (1999-2000) (Table 9.2). Not much variation was observed in the diversity index (H′) in migratory and non-migratory seasons.

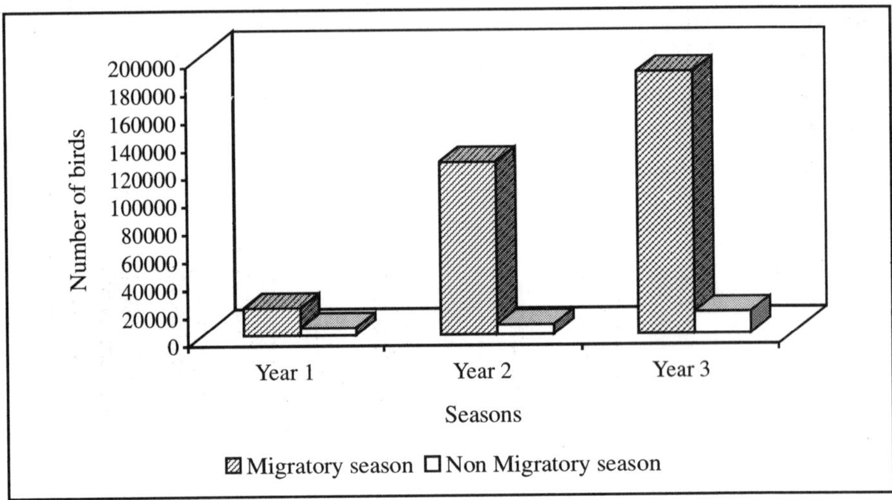

Fig. 9.4: Abundance of birds in the migratory and non-migratory seasons (1998 - 2001)

Table 9.2: Diversity Index (H') during migratory and non-migratory season

Year	Migratory Season	Non-migratory season
Year 1	2.51	2.60
Year 2	2.98	2.07
Year 3	2.87	2.79

Evenness Indices

Evenness Indices during the three migratory seasons were also calculated. Highest Evenness Index was obtained during the second year migratory season (1999-2000) (E1 = 0.62; E2 = 0.16) followed by the third year (2000-2001) (E1 = 0.59; E2 = 0.14) and the first year (1998-1999) (E1 = 0.57; E2 = 0.14). Overall Evenness Index during the migratory season was E1 = 0.60 and E2 = 0.13.

Population fluctuation of waders

Out of 36 species of waders, Whiskered Tern was highest in dominance followed by Little Stint and Wood Sandpiper. Species like Bar-tailed Godwit, Kentish Plover, Pied Avocet, Greater Painted Snipe, Ruddy Turnstone, Common Redshank, Common Snipe, Pintail Snipe, Eurasian Woodcock, Greater Painted-Snipe, Great Knot and Whimbrel showed low dominance index (Table 9.3). The percentage abundance of waders in the intensive study areas is diagrammatically shown in (Figs. 9.5 to 9.6). Whiskered Tern, Wood Sandpiper, Little Stint, Little

Ringed Plover, Curlew Sandpiper, Common Sandpiper and Eurasian Curlew were recorded highest at Kanjany (Fig. 9.5a) and Enamavu (Fig. 9.5b). Whiskered Tern, Wood Sandpiper and Little Ringed Plover were highest at Chettupuzha (Fig. 9.6a) and Parappur (Fig. 9.6b).

Table 9.3: Number of waders recorded in the three migratory seasons (1998-1999, 1999-2000 and 2000–2001)

Sl. No.	Common name	Chettu-puzha	Kanjany	Enamavu	Parappur	Total	Dominance Index
1.	Whiskered Tern	5756	65317	25114	15009	111196	59.8
2.	Little Stint	0	8973	8827	1341	19141	10.3
3.	Wood Sandpiper	1076	10793	5092	1863	18824	10.2
4.	Little Ringed Plover	614	5585	702	603	7504	4.1
5.	Curlew Sandpiper	0	2530	1382	1662	5574	3.0
6.	Common Sandpiper	190	1245	1445	630	3510	1.9
7.	Eurasian Curlew	656	1629	1182	0	3467	1.9
8.	Ruff	0	1933	470	28	2431	1.3
9.	Small Pratincole	0	2231	44	0	2275	1.2
10.	Marsh Sandpiper	526	1562	30	34	2152	1.2
11.	Green Sandpiper	2	215	1166	113	1496	0.8
12.	Lesser Sand Plover	84	362	936	0	1382	0.8
13.	Red-wattled Lapwing	123	472	214	427	1236	0.7
14.	Temminck's Stint	0	256	478	125	859	0.5
15.	Broad-billed Sandpiper	0	25	382	407	814	0.4
16.	Pacific Golden-Plover	32	660	0	0	692	0.4
17.	Caspian Tern	0	630	0	0	630	0.3
18.	Black-winged Stilt	12	291	47	31	381	0.2
19.	Common Greenshank	56	143	22	87	308	0.2
20.	Dunlin	0	0	324	0	324	0.2
21.	Black-tailed Godwit	154	34	36	0	224	0.1
22.	Sanderling	0	198	0	24	222	0.1
23.	Pheasant-tailed Jacana	13	0	148	0	161	0.1
24.	Bronze-winged Jacana	30	1	64	0	95	0.1
25.	Bar-tailed Godwit	0	46	28	0	74	+
26.	Kentish Plover	0	26	6	30	62	+
27.	Pied Avocet	0	58	0	0	58	+
28.	Greater Painted-Snipe	0	52	2	0	54	+
29.	Ruddy Turnstone	0	0	53	0	53	+
30.	Common Redshank	0	43	7	8	58	+
31.	Common Snipe	0	26	2	6	34	+
32.	Pintail Snipe	0	30	0	0	30	+
33.	Eurasian Woodcock	0	0	41	0	41	+
34.	Terek Sandpiper	0	33	2	0	35	+
35.	Great Knot	0	16	0	0	16	+
36.	Whimbrel	0	0	10	0	10	+
	Total	**9324**	**105415**	**48256**	**22428**	**185423**	**100**

(+ = Dominance less than 0.05)

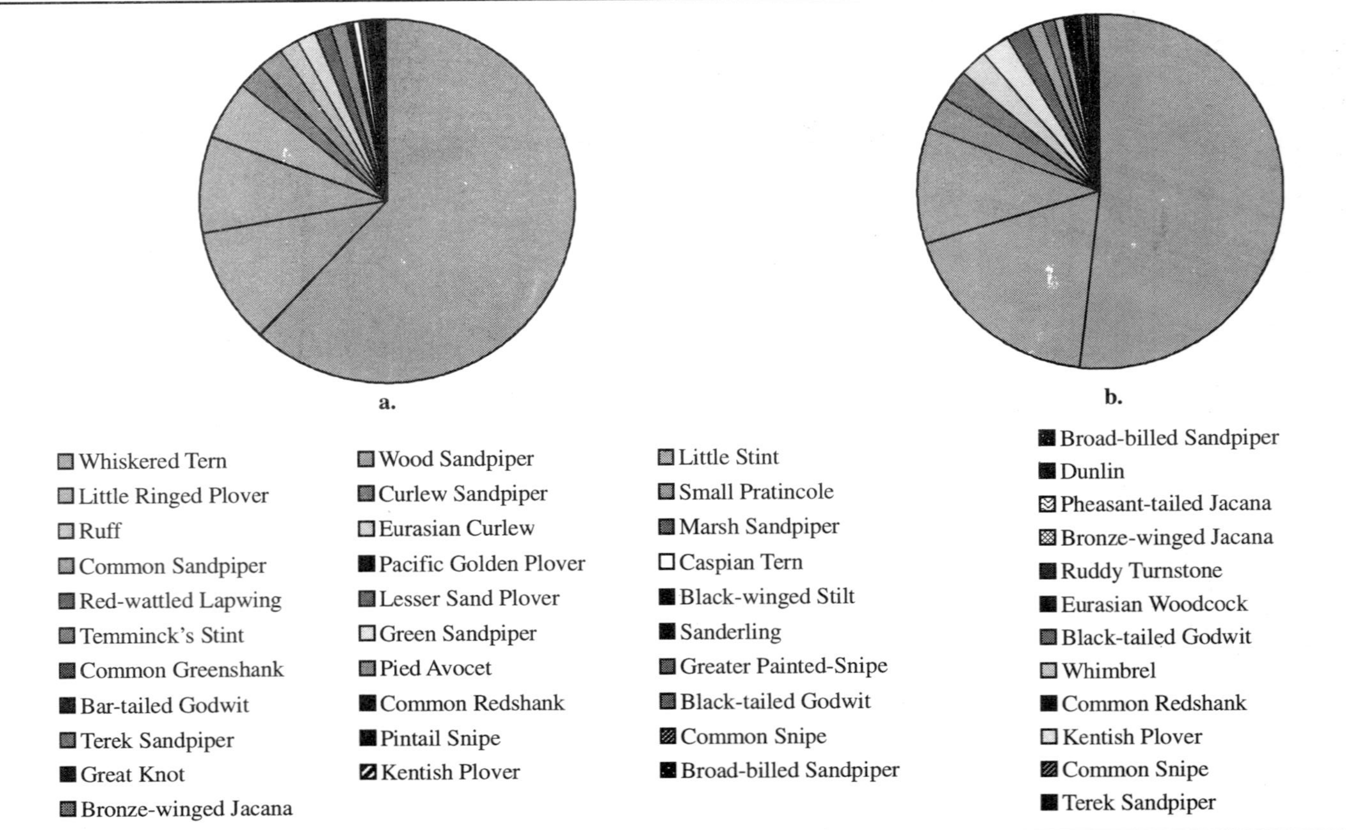

Fig. 9.5: Percentage abundance of waders in Kanjany (a) and Enamavu (b)

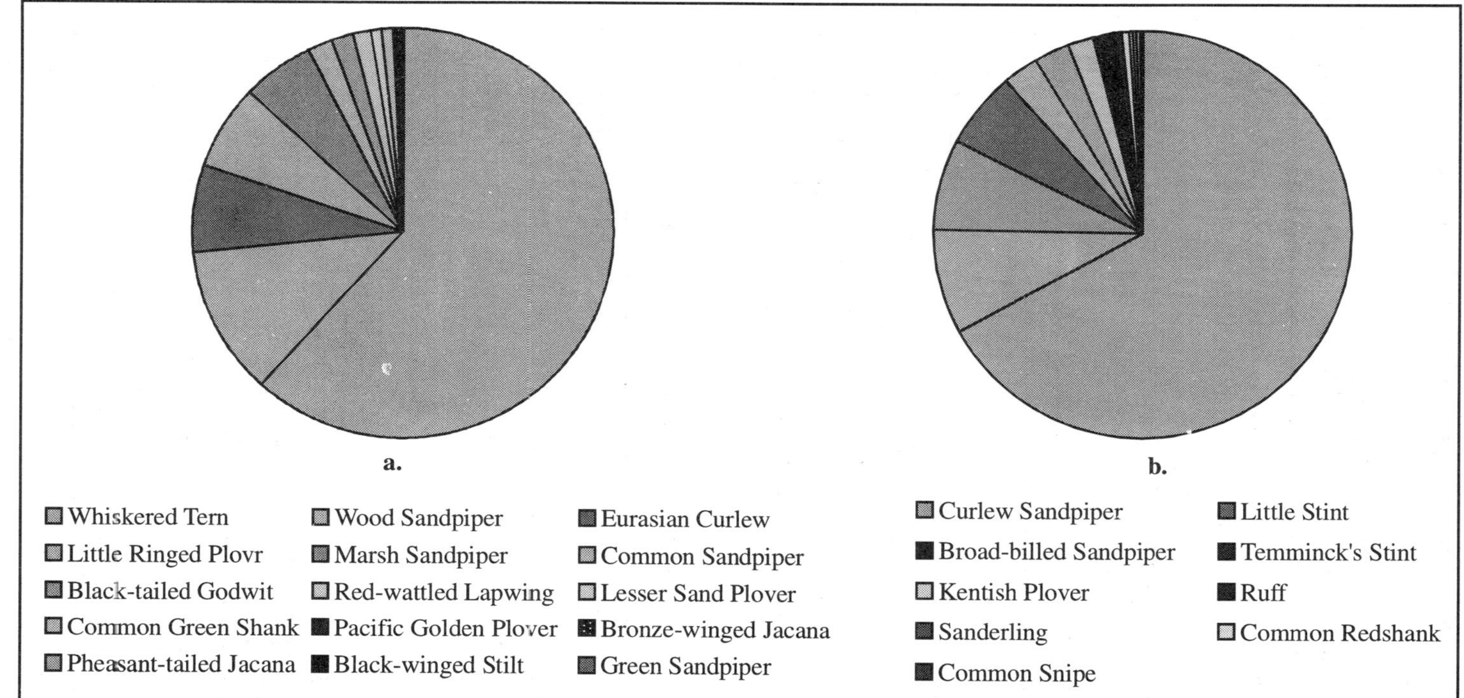

Fig. 9.6: Percentage abundance of waders in Chettupuzha (a) and Parappur (b)

Factors governing the changes in the bird population

Rainfall

A negative relationship was observed between the species richness, abundance and diversity index of birds with total monthly rainfall. When the rainfall increased, species richness and total number of birds decreased. A significant negative correlation was observed between the species richness and the rainfall ($r = -0.65$, $P<0.05$, $n = 12$) (Fig. 9.7). Similarly, there was a significant negative correlation between the rainfall and abundance of birds ($r = -0.50$, $P<0.05$, $n = 12$) (Fig. 9.8) and density of birds ($r = -0.49$, $P<0.10$, $n = 12$) (Fig. 9.9).

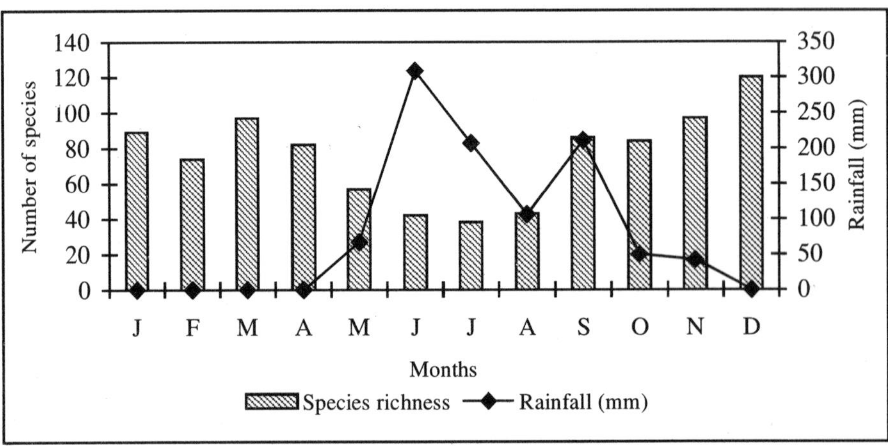

Fig. 9.7: Relationship between species richness and rainfall in the Kole wetlands

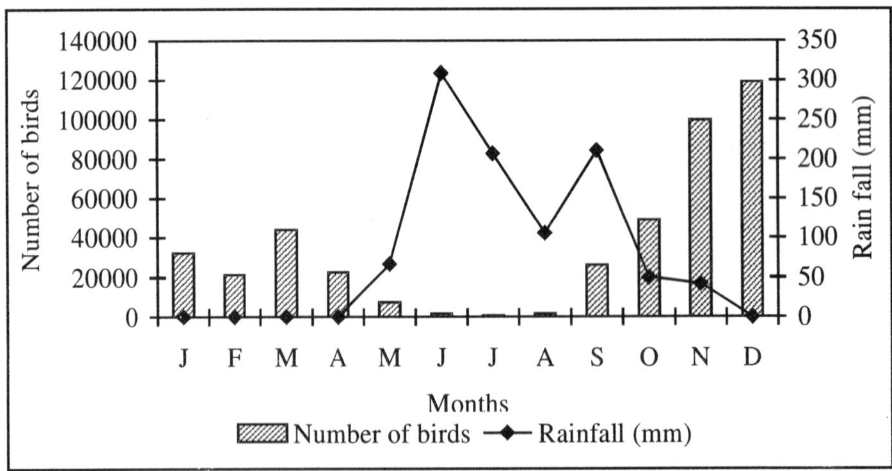

Fig. 9.8: Relationship between abundance of birds and rainfall in the Kole wetlands

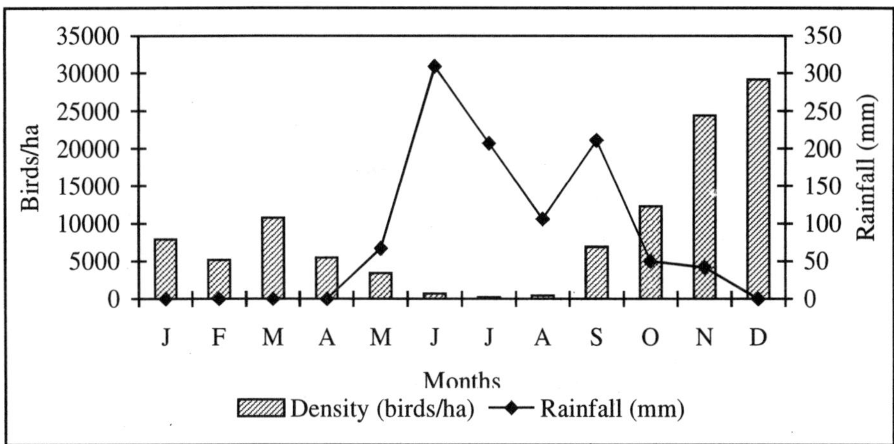

Fig. 9.9: Relationship between density of birds and rainfall in the Kole wetlands

Water depth

Water depth in the Kole wetlands varied from 8 cm in the month of February to 130 cm in July. Species richness decreased as the water depth increased in the wetlands. A significant negative correlation was observed between the water depth and the bird population parameters, (Richness: $r = -0.85$, $P<0.01$, $n = 12$) (Fig. 9.10), (Abundance: $r = -0.60$, $P<0.02$, $n = 12$) (Fig. 9.11), (Density: $r = -0.59$, $P<0.05$, $n = 12$) (Fig. 9.12)

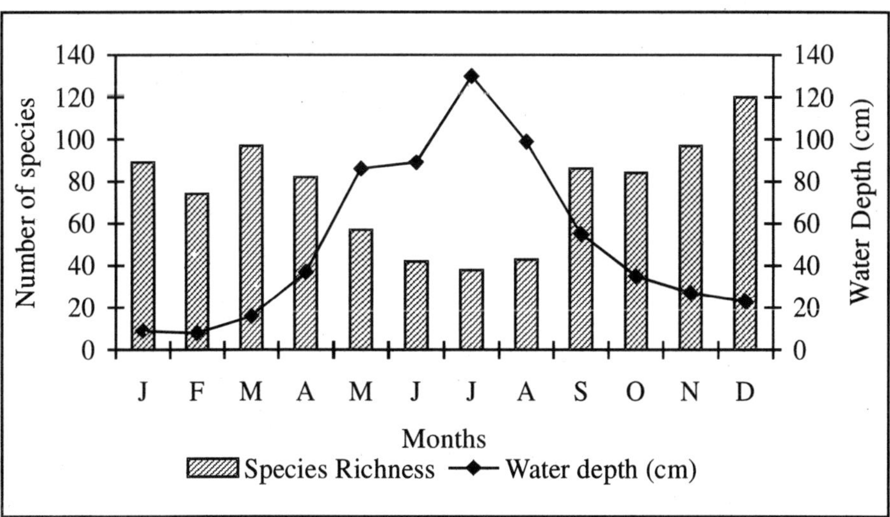

Fig. 9.10: Relationship between species richness and water depth in the Kole wetlands

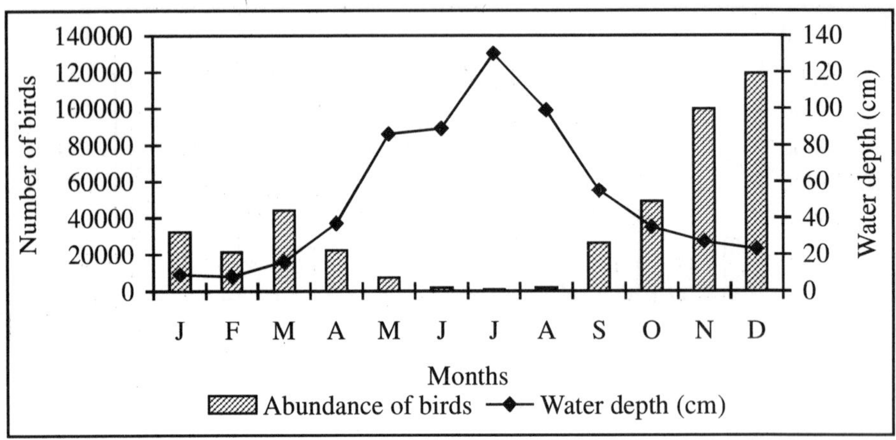

Fig. 9.11: Relationship between abundance and to water depth in the Kole wetlands

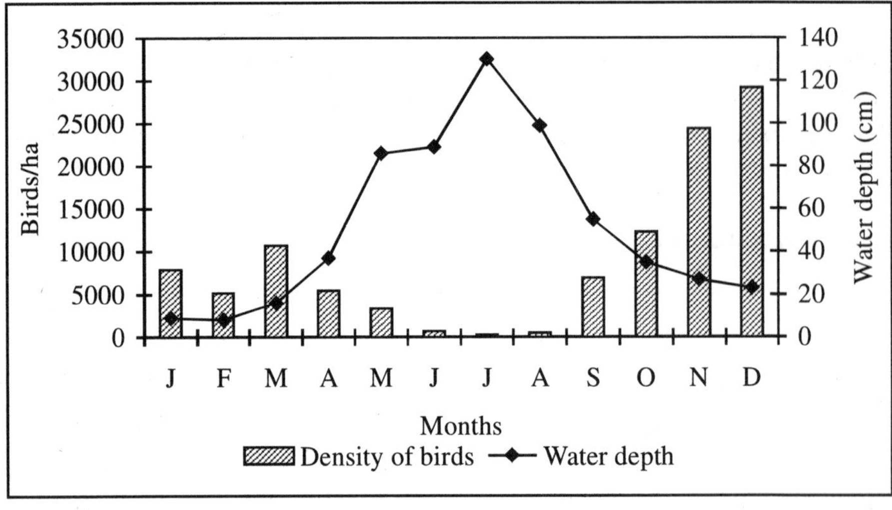

Fig. 9.12: Relationship between density and water depth in the Kole wetlands

Relationship between rainfall and wetland species

Correlation between the rainfall and the water depth with the six Orders of wetland species indicated that significant positive correlation was found between the Order Ciconiiformes and the water depth and the rainfall (Table 9.4). Similarly, significant correlation was found between the Anseriformes and the water depth.

Table 9.4: Relationship between rainfall, water depth and wetland bird Orders in Kole wetlands

S. No.	Order	Correlation coefficient (Rainfall)	Correlation coefficient (Water depth)
1.	Podicipediformes	0.1380 n.s.	0.0525 n.s.
2.	Pelecaniformes	0.3872 n.s.	0.4783 * (P < 0.10, n = 12)
3.	Ciconiiformes	0.5130 ** (P < 0.05,n = 12)	0.5763 ** (P < 0.05, n = 12)
4.	Anseriformes	0.4496 * (P < 0.10, n = 12)	0.4075 * (P < 0.10, n = 12)
5.	Gruiformes	0.1203 n.s.	0.0846 n.s.
6.	Charadriiformes	0.2761 n.s.	0.4500 * (P < 0.10, n= 12)

* = Significant at 10% level; ** = Significant at 5 % level; n.s. = Not significant

Population fluctuations of certain resident species in relation to environmental factors

Cattle Egret (*Bubulcus ibis*)

Monthly variations in the population of Cattle Egret revealed that the population was high in December and no birds were recorded during the southwest monsoon (Fig. 9.13). The variations of Cattle Egret population were negatively correlated with the rainfall ($r = -0.7080$; $P < 0.02$; $n = 12$) and the water depth ($r = -0.7730$; $P < 0.02$; $n = 12$).

Little Egret (*Egretta garzetta*)

The highest population of Little Egret was in November followed by December and October. During the month of June and July, no Little Egret was observed (Fig. 9.13). This was significantly correlated with the water depth ($r = -0.4296$, $P < 0.10$, $n = 12$). No significant correlation was found with the rainfall.

Little Cormorant (*Phalacrocorax niger*)

Little Cormorant was recorded from all the areas of Kole wetlands. Population of Little Cormorant fluctuated every month and it was highest in the month of October followed by November and December

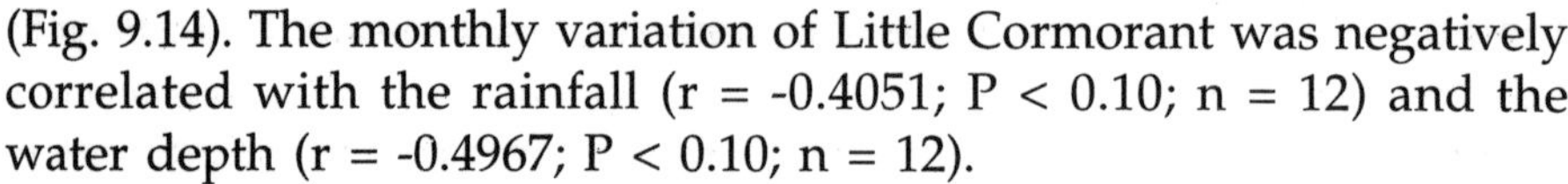

(Fig. 9.14). The monthly variation of Little Cormorant was negatively correlated with the rainfall (r = -0.4051; P < 0.10; n = 12) and the water depth (r = -0.4967; P < 0.10; n = 12).

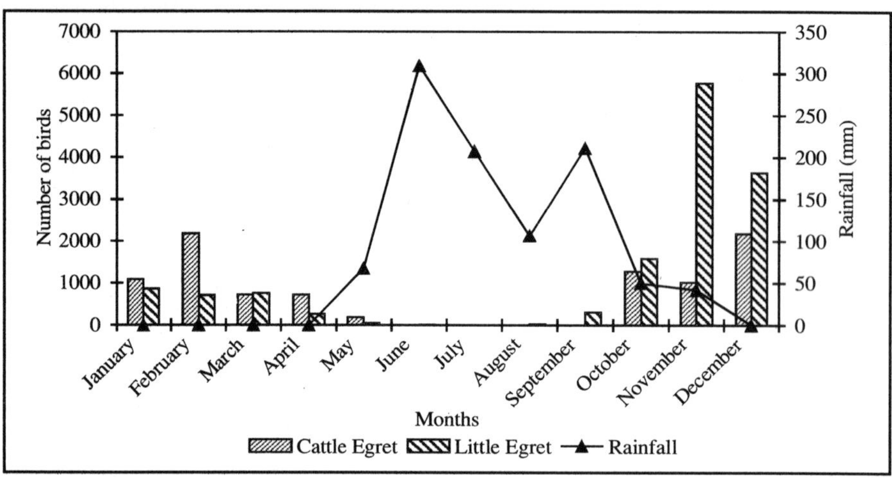

Fig. 9.13: Population dynamics of Cattle Egret and Little Egret in relation to rainfall

Median Egret (*Mesophoyx intermedia*)

Population size of the Median Egret increased from 72 individuals in August to 3,996 individuals in November. The highest population was observed in November followed by December (Fig. 9.14). The monthly variations were not significantly correlated with the rainfall and the water depth.

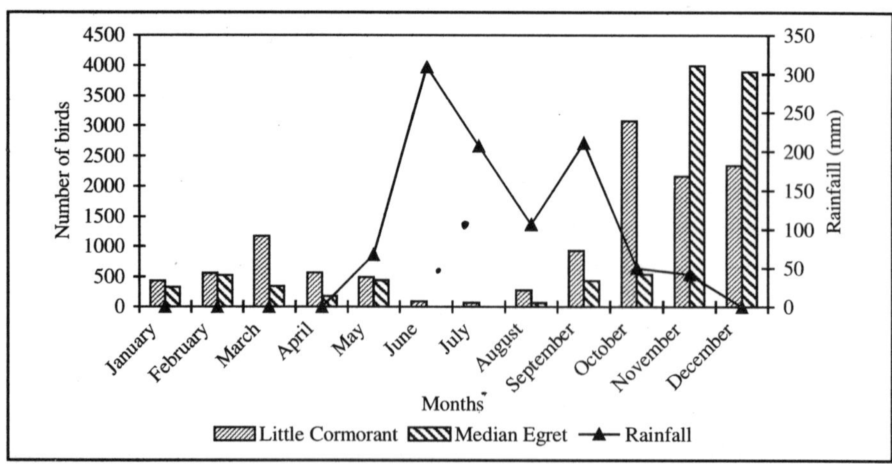

Fig. 9.14: Population dynamics of Little Cormorant and Median Egret in relation to rainfall

Grey Heron (*Ardea cinerea*)

Population of Grey Heron varied from 4 individuals in June to 175 individuals in December. Highest population was recorded in December followed by February (Fig. 9.15). The population had a negative correlation with the rainfall ($r = -0.5110$; $P < 0.02$; $n = 12$) and the water depth ($r = -0.5191$; $P < 0.02$; $n = 12$).

Purple Heron (*Ardea purpurea*)

Highest population of Purple Heron was observed in December, March and July (Fig. 9.15). No Significant correlation was obtained between the abundance of Purple Heron with rainfall and water depth.

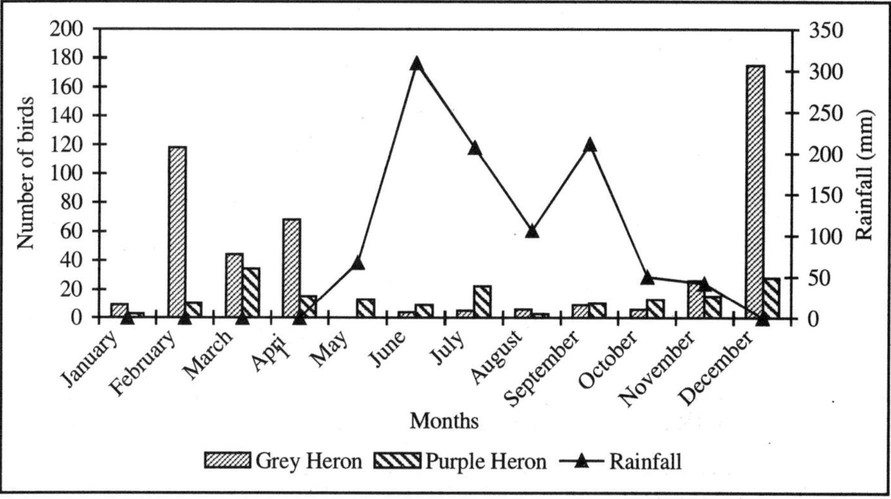

Fig. 9.15: Population dynamics of Grey Heron and Purple Heron in relation to rainfall

Pond Heron (*Ardeola grayii*)

Highest number of Pond Heron was recorded in December followed by November and the species was absent during southwest monsoon (Fig. 9.16). Population of Pond Heron was negatively correlated with the rainfall ($r = -0.5274$; $P < 0.05$; $n = 12$) and the water depth ($r = -0.6073$; $P < 0.02$; $n = 12$).

Population variation of birds in relation to paddy growth

The abundance of birds varied in relation to the height of paddy. Highest number of birds was recorded during the replanting season and sowing period (0 to 11 cm) and there was a decline in the bird population, when the paddy reached a height of 21 to 30 cm. Sudden influx in the bird population was recorded, when paddy reached a height of 31 to 40 cm height during the flowering season (Fig. 9.17). Flowering season attracted many nectarivores species. Weaverbirds and Munias were congregated in huge numbers during the harvesting period of paddy. A significant negative correlation was found between the paddy height and the abundance of birds (r = -0.7898; P<0.01; n=8).

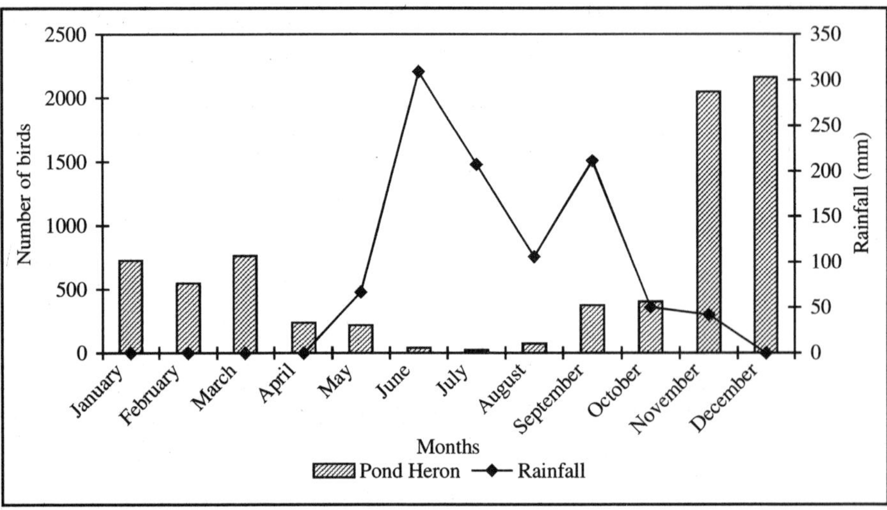

Fig. 9.16: Population dynamics of Pond Heron in relation to rainfall

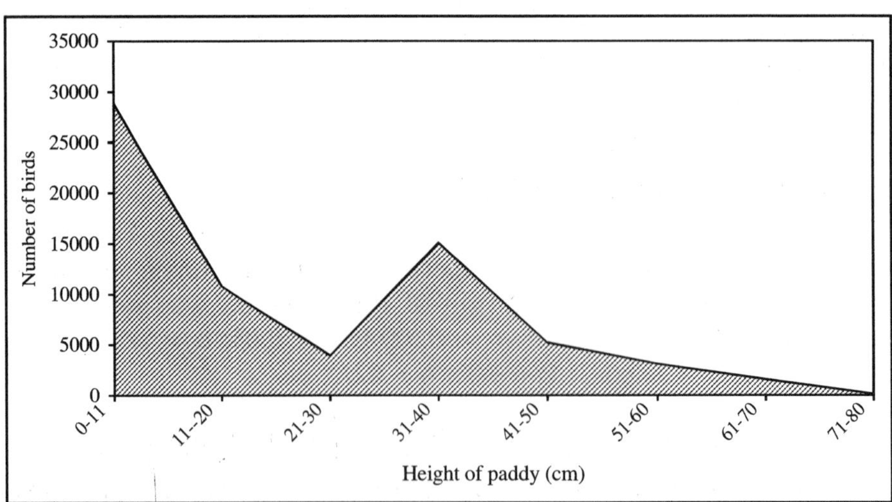

Fig. 9.17: Relationship between abundance of birds and height of paddy

Seasonal fluctuation in the abundance of a species is considered as an adaptive phenomenon evolved through ages to derive maximum advantage from the ambient environmental conditions (Koen, 1992). The species composition of birds varied depending on the season of the year. Availability of microhabitats and various food resources were the determining factors, which controlled the seasonal changes of bird species composition. Species richness and abundance of waders were highest during the migratory season. Increase in the size of bird population in the second and third year could be due to the early exposure of the preferred microhabitat of waders. The peak population of birds was during the migratory season and a decrease was recorded during the non-migratory season, which is comparable with earlier studies (Saikia and Bhattacharjee, 1990b; Nagarajan and Thiyagesan, 1996; Acharya, 2000).

Presence of 39 species of waders in the Kole wetlands showed the importance of the wetlands for the trans-continental migratory species. Among the four study sites, highest number of Wood Sandpiper was recorded from Kanjany. This is comparable to the report of Wood Sandpiper from Gulf of Mannar Marine National Park in small numbers by Balachandran (1995). Abundance of Black-tailed Godwit, Great Knot and Marsh Sandpiper were higher in the Kole wetlands, which is compared to other wetlands of Kerala and other States (Hoffmann, 1983; Balachandran, 1995; Acharya and Kar, 1996; Verma *et al.*, 2002).

During the monsoon months, there was reduction in the species richness and abundance of birds at Kole wetlands. Distribution and abundance of birds in the Kole wetlands are influenced simultaneously by habitat preferences, seasonal and geographic variation in habitat availability and variation in the climatic conditions. The monthly variations of richness, abundance, density and diversity of bird population were correlated with the rainfall and the water depth, which showed that the rainfall and water depth were the major factors influencing the abundance of birds at Kole wetlands. When the water depth and the rainfall increased, number of birds decreased. There was a strong response to water depth in the Kole wetlands with considerable decrease of population in the wet seasons. The number of birds during November, December and January were highest, when precipitation was less and extensive mud flats and shallow waters were available in the Kole wetlands. It is clear that the environmental factors are determining the wetland bird community primarily by their direct or indirect impact on the availability and abundance of birds prey. The role of abundance of food on wetland bird density have been well established (Sjoberg, 1989; Parker *et al.*, 1992; Nagarajan and Thiyagesan, 1996) which agree with the present study.

According to Kushlan (1986), wading birds may use diverse strategies to cope up with the seasonal functions of water depth. Hafner and Britton (1983) reported that the food availability between different habitats played a sensitive role in the distribution and seasonal movement. The distribution of wetland birds at Kole wetlands is thus due to the local depletion of food resources mediated by increased interference of other birds, which is comparable with the reports of Goss-Custard (1977) and Zwarts (1978). The water depth was another important factor, which could be used to predict the wetland bird population of Kole wetlands. A relationship between the habitat use of birds and water depth was reported from the Great Vedaranyam Salt Swamp (Sampath and Krishnamurthy, 1989).

The Kole wetlands served many avian species for a wide variety of purposes such as nesting and roosting and act as wintering ground. Maintenance of habitat diversity is essential for avian diversity, which requires various natural water regimes and plant communities. The present study showed that the Kole wetlands are one of the important regions in Kerala for winter visitors. During the migratory season, the Kole wetlands supported more waders than other sites in Kerala.

❑❑❑

Chapter 10

Habitat Utilisation of Wetland Birds

Habitat is the most valuable variable in ornithological studies as it provides food and cover essential for the survival of the population (White and Garrott, 1985). Habitat is an area containing or influenced by the suitable resources and environmental conditions that determine the presence, survival and reproduction of a population (Morrison *et al.*, 1998). Cody (1985) described habitat as areas where an organism's need for food and shelter are met. Habitat selection and use by birds is fundamental in understanding their biology and management and it has been one of the most well studied aspects of avian biology, particularly as habitats are altered by human activities (Pain and Pienkowski, 1996).

According to the optimal foraging theory, animals may use the areas offering higher food extensively (Stephens and Krebs, 1987). By using the most productive patches, birds may have a high ingestion rate, allowing to reduce the time exposed to predators and having more time to other activities. However, some authors have obtained contradictory results also. They have not found clear relationship between bird abundance and food availability (Wiens, 1984 and Herrera, 1988). The ability to exploit food resources is one of the most frequently suggested reasons for habitat selection by birds (Caraco, 1980) and would be expected to be primary basis for selection of foraging habitat.

Counting of a particular bird species in a study area is much valuable, when it is related to habitat that is responsible for bird occurrence or abundance (Cody, 1985 and Wiens, 1989). Many wetland species migrate long distances to spend the non-breeding period in wetland habitats ranging from north temperate to south temperate regions of the world. During this period, they relay on a few, key stopover sites to complete their annual migratory cycle. These sites often provide unique combination of food resources, safe roosting habitat and less disturbance, which are the key factors necessary to support large number of birds. Waders are particularly at risk from habitat deterioration of migration and wintering sites because of their tendency to concentrate at a few sites at precise times (Myers *et al.*, 1987). A common approach to understanding the relationships of bird species to their habitat is to examine the relationship of a single bird species to specific vegetative characteristic found in an individual's territory (Shugart and Urban, 1986). However, it is evident that bird population is limited not only by the individual patches occupied, but also by the surrounding landscape (Whitecomb *et al.*, 1981). Birds may select different habitat features from a number of different spatial scales (Steele, 1992).

METHODS

Microhabitat utilisation of wetland birds was studied by recording the microhabitat of the bird at the first sight. Observations were made after sunrise up to 1000 hour. Plant species from the Kole wetlands were collected and identified with the help of plant taxonomists.

Categories of microhabitat

Microhabitats were classified into categories as given below.

Floating vegetation (FV) : Open water with floating species such as *Eichhornea crassipes*, *Pistia* sp. Usually found in all the areas during monsoon and restricted to canals during summer months.

Bund (BU) : Bunds erected mainly by earth and stones separate the canals and the paddy fields. During the monsoon months, bunds were the only structure above water. Different types of bunds were considered together for collecting the data.

Bamboo pole (BP) : Bamboo poles were used for constructing the temporary bridges over the canal. Some bamboo poles were pitched in the paddy fields for demonstrating scaring devices like colored plastic bags.

Trees (TR) : Some small and medium sized trees were seen on the bunds and the species *Alstonia scholaris* was common in the area.

Electric line (EL) : Electric line of different voltages (11 KV to 400 KV), which pass through the wetlands is considered as a microhabitat.

Water edge (WE) : Edges of water bodies like canals and paddy fields come under this category.

Shallow water (SH) : Shallow water with a depth up to 20 cm usually free from the aquatic vegetation is a microhabitat. It usually emerged during summer months, when the water is pumped to the canals for paddy cultivation. This microhabitat was found only in the summer months.

Mudflat (MF) : Soft mud and water with a depth up to 10 cm was another microhabitat of the area. It was usually found in the paddy fields before replanting of the paddy.

Grass (GR) : The Kole wetlands had patches of grass with moisture content and are usually seen on the bunds. *Arundinaria donax* a tall grass was found in the area, which attained a height of 180 cm to 240 cm. Farmers burned this during summer months.

Paddy field (PF) : Paddy cultivated areas in different stages *viz.* sown, replanted, flowering and panicle stages were considered as a microhabitat.

Open water (OW) : Open water is defined as the vast expanse of water without vegetation. This was found mostly during monsoon.

Harvested paddy field (HPF) : An area where paddy was harvested and the land ploughed later is another microhabitat.

Rarefaction

The number of species invariably increases with sample size and sampling effort. To cope with this problem, Rarefaction technique is useful for calculating the number of species expected in a sample of n individuals drawn from a population total of N individuals distributed among S species is

$$E(S) = \sum \left\{ 1 - \left[\binom{N - n_i}{n} \Big/ \binom{N}{n} \right] \right\}$$

where n_i = number of individuals in the ith species

The Rarefaction was calculated using the software BioDiversity (Lambshead *et al.*, 1997).

Species richness, diversity and density in different microhabitats

The species richness, diversity and density of birds in different microhabitats were calculated from the daily census of birds.

Niche breadth and overlap indices

Two Niche breadths were calculated namely the Levins and Hulbert using the software STATECOL (Ludwig and Reynolds, 1988). Niche breadth is another method, which mathematically, falls into synonymy with other concepts proposed by Levins (1968). A widely used measure of niche breadth is Levins measure (B) and Levins standardised niche breadth (B′). The following formula was used to estimate niche breadths.

Levins Measure (Levins, 1968)

$$B = 1/\sum_i pi^2$$

where,

B = Levins measure of niche breadth

pi = Proportion, out of all those resources use by the population

Hulbert (1978)

$$B' = \frac{(B-1)}{(n-1)}$$

B′ = Hulbert niche breadth

B = Levins measure of niche breadth

n = number of possible resource states

Overlap indices

Niche overlap indices were calculated to find out the overlap in microhabitat use. The equations are based on Petraitis (1979), which is further described in Ludwig and Reynolds (1988). The overlap indices were calculated using the program STATECOL (Ludwig and Reynolds, 1988).

General Overlap GO = e^E

$$where\ E = \frac{\Sigma_i\ ^s E\ ^r_j [n_{ij}\ (C_j\text{-In } P_{ij}]}{T}$$

The Statistic V = -2T ln GO

Minimum GO (GO_{mim}) = e (1/T) [Σ I-1 (N_i In N_i) – (T In T)]

$$\text{Adjusted GO } (GO_{adj}) = \frac{GO - GO_{min}}{1 - GO_{min}}$$

Habitat utilisation of selected species

The utilisation of microhabitats by bird species was determined by regular observations. Species on which at least 50 observations were recorded alone was included in the analysis and the sample size varied from 60 to 371. The mean value for each variable was calculated from the combined data of 36 months of the study for respective microhabitat type.

Microhabitat use in different seasons

Analysis of variance (ANOVA) (Gomez and Gomez, 1984) and Student Newman Keuls test (SNK) (Montogomery, 1991) were performed to know the changes in the habitat and season on the variation of bird species richness, diversity and density using the SPSS 10.0. ANOVA was carried out using log-transformed data.

Plant species

A total of twenty-three plant species were identified from the Kole wetlands, which belong to thirteen families (Table 10.1).

Table 10.1: Plant species identified from Kole wetlands

Sl. No.	Family	Species
1.	Fabaceae	*Smithia* sp.
2.	Onagraceae	*Ludwigia perennis*
3.	Rubiaceae	*Hedyotis* sp.
4.	Asteraceae	*Sphaeranthus indicus*
5.	Lobeliaceae	*Lobelia dichotoma*
6.	Hydrophyllaceae	*Hydrolea zeylanica*
7.	Convolvulaceae	*Ipomoea aquatica*
8.	Scrophulariaceae	*Limnophila heterophylla*
9.	Acanthaceae	*Justicia japonica*
10.	Pontederiaceae	*Eichhornea crassipes*
11.	Commelinaceae	*Cyanotis* sp.
12.	Cyperaceae	*Bulbostylis* sp.
13.		*Carex* sp.
14.		*Fimbristylis* sp.
15.		*Kyllinga bulbosa*
16.		*Lipocarpha chinensis*
17.		*Mariscus* sp.
18.		*Pycreus* sp.
19.		*Rhynchospora corymbosa*
20.	Poaceae	*Coix lachryma-Jobi*
21.		*Heteropogon contortus*
22.		*Imperata cylindrica*
23.		*Paspalum scrobiculatum*

Rarefaction

A comparison of number of species in each microhabitat for fixed number of individuals encountered is a useful exercise. For this purpose, expected number of species was calculated for varying number of individuals and presented in Fig. 10.1 using rarefaction formula. The data shows that electric line microhabitat had more number of species followed by mud flats and shallow water.

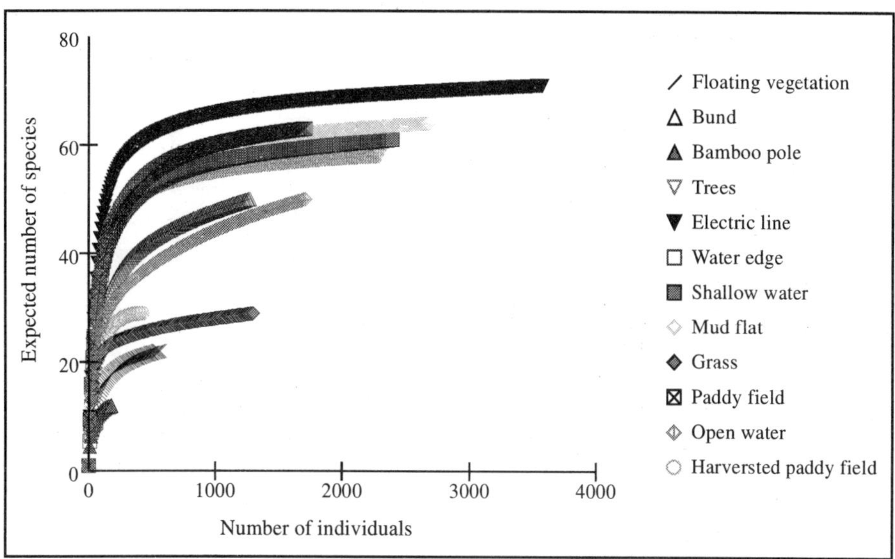

Fig. 10.1: Rarefaction for bird species in different microhabitats

Species richness, diversity and density in different microhabitats

Species richness of birds was highest in the electric line microhabitat (74) followed by mud flats (69). Species diversity index (H′) was highest on trees (3.44) and lowest on bund (1.38) (Fig. 10.2). Density of birds was highest in shallow water followed by paddy fields and mud flats (Fig. 10.3).

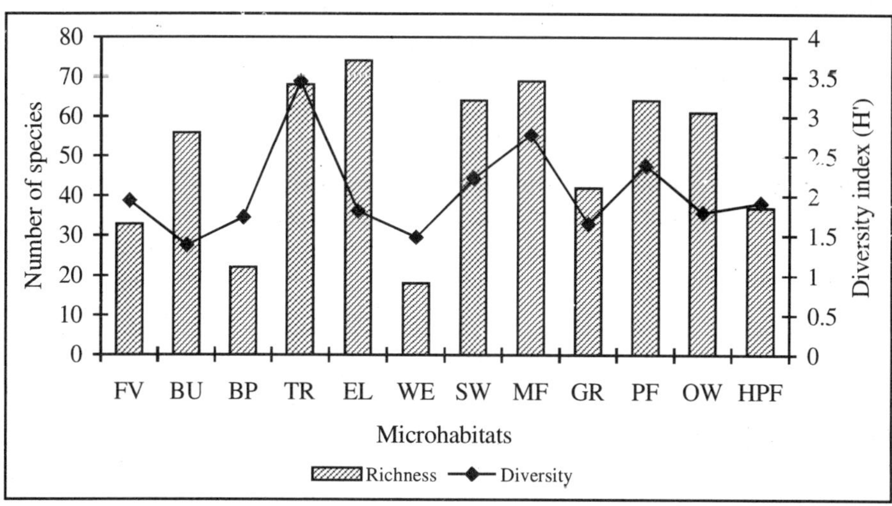

Fig. 10.2: Bird species richness and diversity index (H’) in different microhabitats in the Kole wetlands

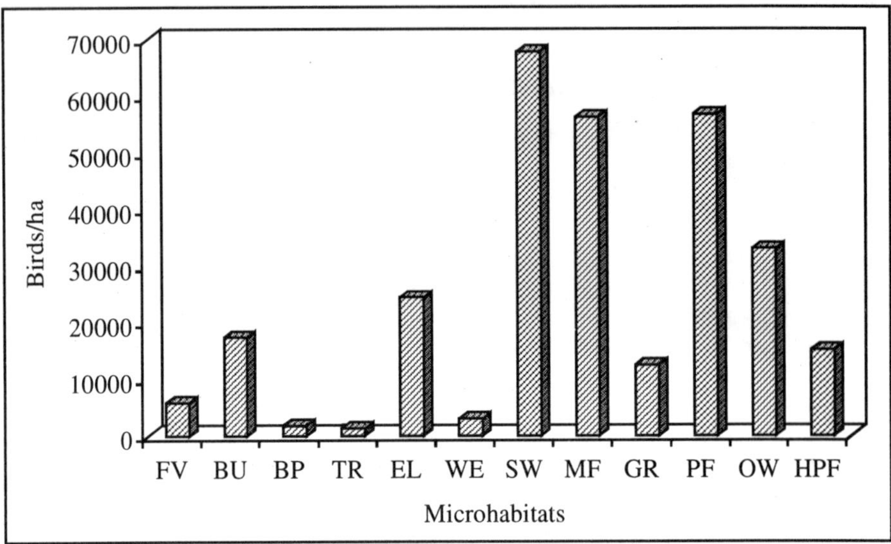

Fig. 10.3: Density of birds in different microhabitats in the Kole wetlands

FV = Floating vegetation, BU = Bund, BP = Bamboo pole, TR = Trees, EL =Electric line, WE = Water edge, SW = Shallow water, MF = Mud flats, GR = Grass, PF = Paddy fields, OW = Open water, HPF = Harvested paddy fields.

Niche breadth and overlap indices

The general overlap of microhabitats between 11 species was only fifty per cent (GO = 0.534; V = 1377.61) with the critical value of chi-square and df = 110, P = 0.05 (Table 10.2). It showed that there is no complete overlap between the microhabitat utilization of different bird species. Highest niche breadth was recorded for Red-wattled Lapwing followed by Indian Pond-Heron and Little Cormorant (Table 10.3).

Table 10.2: General overlap index values of different microhabitat resources for general (GO), minimum GO (Gmin) and adjusted GO (Gadj)

Number of Species	Indices				
	GO	Gmin	Gadj	V	df
11	0.534	0.091	0.488	1377.61	110

Habitat utilisation of selected bird species

Habitat utilisation pattern of 11 selected species of birds showed that except waders, all the wetland species utilised all the available microhabitats (Table 10.3).

The shallow water was highly preferred among the twelve microhabitats. The utilisation pattern of microhabitats differed from species to species. Little Egrets were utilising all the microhabitats except grass and the species frequented shallow water more followed by paddy fields. Observations indicated that Cattle Egret highly preferred paddy fields and the species had a characteristic affinity for the paddy fields than other habitats. Indian Pond-Heron preferred the paddy fields and the shallow water. Cattle Egret and Indian Pond-Heron highly preferred the paddy fields than other microhabitats. Red-wattled Lapwing's preferred habitat was mud flats.

Large Egret preferred the paddy fields more than other microhabitats. Similarly, Common Sandpiper preferred the mud flats and shallow water. Wood Sandpiper was observed in the mud flats followed by the shallow waters. Black-winged Stilt preferred the shallow waters and Whiskered Tern preferred the shallow waters and mud flats. Marsh Sandpiper preferred the shallow water and mud flats for foraging. Median Egret preferred the shallow water and the paddy fields. Habitat overlapping was also observed among birds in the Kole wetlands. Most of the species like Little Egret, Median Egret, Black-winged Stilt, Marsh Sandpiper and Common Sandpiper mainly used the shallow water and the mud flats, and other species like waders preferred single habitat. The distribution pattern of bird community based on the utilisation of different microhabitat is depicted in Fig. 10.4.

Influence of microhabitat and seasons on bird community parameters

Two-way analysis of variance (ANOVA) was performed on the bird species richness, diversity and density to test the significant difference due to the habitats and seasons (dry, wet-I, and wet-II).

Species richness: Species richness showed significant difference between the habitats and between the seasons. Similarly, interaction between the habitats and seasons also showed statistically significant difference (Table 10.4).

Species diversity Index (H′): Species diversity index (H′) showed significant difference between the habitats and between the seasons and interaction between the habitat and seasons showed significant difference (Table 10.4).

Fig. 10.4: Feeding and perching sites of birds in the Kole wetlands
1. Waterfowls, 2. Little Ringed Plover, 3. Little Stint, 4. Snipe, 5. Black-winged Stilt, 6. Pond Heron, 7. Cattle Egret, 8. Little Egret, 9. Small Bee-eater, 10. Swallow, 11. Purple Moorhen, 12. Pheasant-tailed Jacana, 13. Ruddy-breasted Crake, 14. Slaty-legged Crake, 15. White-breasted Waterhen, 16. Blue Rock Pigeon, 17. Streaked Fantail-Warbler, 18. Red Munia, 19. Blyth's Reed-Warbler, 20. Munia, 21. Indian Great Reed-Warbler, 22. Baya Weaver, 23. Rose-ringed Parakeet, 24. Black Drongo, 25. Myna, 26. Shikra, 27. White-breasted Kingfisher, 28. Marsh Harrier, 29. Kite, 30. Whiskered Tern.

Table 10.3: Habitat utilisation of birds in the Kole wetlands (Percentage)

Name of the species	Floating vegetation	Bund	Bamboo pole	Trees	Electric line	Water edge	Shallow water	Mud flat	Grass	Paddy field	Open Water	Harvested paddy field	Total no. of sighting (n)	Niche breadth	
														Levins (B)	Hulbert (B')
Little Egret	0.14	2.93	0.32	0.03	0.03	1.52	44.43	17.17	0	26.59	0.21	6.62	261	3.29	0.22
Little Ccrmorant	3.76	8.01	7.61	0.97	2.66	0.87	11.87	15.59	0	1.35	43.02	4.2	326	4.17	0.32
Indian Pond-Heron	7.54	11.24	0.43	0.54	0.15	0.12	16.92	12.22	0.01	40.86	1.43	8.53	371	4.23	0.29
Cattle Egret	0.69	0.9	0.05	0.05	0	0.54	8.73	4.4	0.03	52.35	0.18	32.09	214	2.58	0.16
Whiskered Tern	0.02	0.09	0.57	0.15	24.12	0	43.87	17.86	0	8.42	3.53	2.38	232	3.43	0.27
Lesser Whistling-Duck	76.12	1.37	0	0	0.21	0	3.77	0	0.1	0	18.05	0.38	58	1.62	0.10
Median Egret	0.8	1.24	0	0	0.02	0.66	41.35	4.6	0	47.94	0.01	3.39	122	2.47	0.18
Common Sandpiper	0.39	7.32	0	0	0	0.32	18.41	56.48	0.39	12.96	0	3.73	104	2.65	0.24
Wood Sandpiper	0.01	1.75	0	0	0	0.13	19.42	46.56	0	28.2	0	3.92	152	2.98	0.33
Red-wattled Lapwing	1.06	26.31	0	0.77	1.55	0	9.48	29.79	0.19	17.7	7.74	5.13	170	4.83	0.43
Purple Moorhen	46.53	0	0.05	0.04	0	0.25	27.24	0.04	0.52	0.15	25.19	0	60	2.82	0.23

Density of birds : Density of birds showed significant difference between the habitats and the seasons. Similarly, interaction between the habitat and seasons also showed significant difference (Table 10.4).

Two-way ANOVA was performed on the bird species richness, species diversity index and density of birds to find out significant difference between migratory and non-migratory seasons and habitats.

Species richness : Species richness showed significant difference between the habitats and between the migratory and non-migratory season. Similarly, interaction between the habitat and seasons also showed significant difference (Table 10.5).

Species diversity Index (H') : Species diversity Index (H') showed significant difference between the habitats and between the seasons and interaction between the habitat and seasons also showed significant difference (Table 10.5).

Density of birds : Density of birds showed significant difference between the habitats and between the seasons. Similarly, interaction between the habitats and seasons also showed significant difference (Table 10.5).

Microhabitat use in different seasons

The use of microhabitat by birds was compared using Student Newman Keuls test. Species richness, diversity and density varied in different seasons in different microhabitats. Horizontal line connects the subsets of habitat with reference to the corresponding variable. (Table 10.6 to 10.10)

Dry Season: Mean species richness was significantly higher on electric line microhabitat followed by paddy fields and shallow water. Diversity was higher on trees followed by shallow water. Density of birds was significantly higher in the paddy fields followed by open water and mud flats (Table 10.6)

Wet-I season: Mean species richness was significantly higher on electric lines followed by bunds and floating vegetation. Diversity was higher on electric lines followed by trees. Density of birds was significantly higher in floating vegetation followed by open water and electric line (Table 10.7).

Table 10.4: Analysis of Variance for bird community parameters (Dry, Wet-I and Wet-II)

Sources of Variation	DF	Species Richness			Species diversity Index			Density of birds		
		SS	MSS	F	SS	MSS	F	SS	MSS	F
Habitats	11	40.37	3.67	8.56**	12.18	1.11	7.29**	115.63	10.51	3.96**
Seasons	2	28.03	14.01	32.69**	2.11	1.06	6.97*	200.84	100.42	37.86**
Habitats * Seasons	22	27.02	1.23	2.87**	9.37	0.43	2.81**	178.40	8.11	3.06**
Error	72	30.87	0.43		10.92	0.15		190.98	2.65	
Total	**107**	**126.28**			**34.58**			**685.86**		

** P = <0.001; * P = <0.05

Table 10.5: Analysis of Variance for bird community parameters (Migratory and Non-migratory)

Sources of Variation	DF	Species Richness			Species diversity Index			Density of birds		
		SS	MSS	F	SS	MSS	F	SS	MSS	F
Habitats	11	20.87	1.89	9.10**	9.27	0.84	5.91**	81.91	7.45	3.76**
Seasons	1	9.20	9.20	44.13**	1.99	1.99	13.93**	62.83	62.83	31.71**
Habitats * Seasons	11	7.95	0.722	3.47**	3.58	0.33	2.28*	46.49	4.23	2.13*
Error	48	10.00	0.21		6.85	0.14		95.11	1.98	
Total	**71**	**48.03**			**21.69**			**286.34**		

** P = <0.001; * P = <0.05

Table 10.6: Microhabitat use of wetland bird species in the Kole wetlands during dry season (Newman-Keuls test) [a, b]

Variables	Microhabitat* (Mean values in increasing order) Dry season											
	3	**6**	**1**	**12**	**11**	**9**	**2**	**4**	**8**	**7**	**10**	**5**
Species Richness	7.67	8.67	10.67	14.67	17.33	17.67	24.33	28.00	28.33	30.67	31.67	35.33
	±	±	±	±	±	±	±	±	±	±	±	±
	0.33	2.90	2.40	1.45	6.01	2.96	3.33	4.93	3.18	4.11	3.85	2.18
	3	**6**	**12**	**9**	**11**	**2**	**1**	**5**	**10**	**8**	**7**	**4**
Diversity	0.80	1.28	1.21	1.33	1.34	1.46	1.54	1.71	2.06	2.09	2.26	2.72
	±	±	±	±	±	±	±	±	±	±	±	±
	0.08	0.05	0.34	0.18	0.24	0.35	0.26	0.20	0.05	0.07	0.12	0.15
	4	**3**	**1**	**6**	**12**	**9**	**2**	**5**	**7**	**8**	**11**	**10**
Density (Number/ha)	215.33	190.33	452.33	831.00	2947.67	3831.00	3452.33	4796.00	9991.00	7743.00	9381.33	12707.33
	±	±	±	±	±	±	±	±	±	±	±	±
	82.31	22.25	53.82	206.18	1642.16	1991.58	1085.77	373.07	5974.23	3027.52	3832.22	4487.28

*1 = Floating vegetation; 2 = Bund; 3 = Bamboo pole; 4 = Tree; 5 = Electric line; 6 = Water edge; 7 = Shallow water; 8 = Mud flat; 9 = Grass; 10 = Paddy field; 11 = Open water; 12 = Harvested paddy field,
a = Newman-Keuls test was carried out using log transformed data, b = The mean values given above are in original scale

Table 10.7: Microhabitat use of wetland bird species in the Kole wetlands during Wet 1 season (Newman-Keuls test) [a, b]

Variables	Microhabitat* (Mean values in increasing order) Wet-I season											
	10	**8**	**6**	**3**	**12**	**7**	**9**	**11**	**4**	**1**	**2**	**5**
Species Richness	0.00 ± 0.00	0.67 ± 0.51	2.33 ± 0.33	3.00 ± 0.99	3.00 ± 0.57	8.00 ± 3.05	8.67 ± 1.76	13.33 ± 6.56	11.33 ± 2.84	11.33 ± 3.21	14.00 ± 3.78	23.67 ± 1.20
	6	**3**	**8**	**12**	**10**	**11**	**1**	**9**	**7**	**2**	**4**	**5**
Diversity	0.25 ± 0.06	0.52 ± 0.19	0.18 ± 0.17	0.57 ± 0.28	0.00 ± 0.00	1.19 ± 0.14	1.149 ± 0.05	1.53 ± 0.05	1.58 ± 0.26	1.65 ± 0.23	2.12 ± 0.25	2.51 ± 0.17
	10	**8**	**3**	**4**	**6**	**2**	**12**	**9**	**7**	**5**	**11**	**1**
Density (Number/ha)	3.00 ± 3.00	12.33 ± 12.33	52.33 ± 23.70	39.00 ± 13.11	78.00 ± 21.54	296.33 ± 184.20	171.33 ± 70.13	125.67 ± 9.38	511.67 ± 428.64	338.33 ± 107.51	570.00 ± 167.29	1219.67 ± 709.67

*1 = Floating vegetation; 2 = Bund; 3 = Bamboo pole; 4 = Tree; 5 = Electric line; 6 = Water edge; 7 = Shallow water; 8 = Mud flat; 9 = Grass; 10 = Paddy field; 11 = Open water; 12 = Harvested paddy field
a = Newman-Keuls test was carried out using log transformed data, b= The mean values given above are in original scale

Wet-II season : Species richness was significantly higher in mud flats followed by electric lines and shallow water. Diversity was higher on trees and the shallow water. Density of birds was significantly higher in the mud flats followed by shallow water and electric line (Table 10.8).

Migratory season : Species richness was significantly higher in the mud flats followed by electric lines and paddy fields. Diversity was higher on trees, mud flats and shallow water. Density of birds was significantly higher in paddy fields followed by mud flats and shallow water (Table 10.9).

Non-migratory season : Species richness was significantly higher on electric line and on bunds. Diversity was higher on trees, electric lines and shallow water. Density of birds was significantly higher in open water followed by floating vegetation and electric line (Table 10.10).

Among the microhabitats identified, species richness was highest on the electric line and in the mud flats. Species diversity index was higher on the tree and density of birds in the shallow water followed by the paddy fields and the mud flats. The variation in species richness, diversity index and density of birds in different microhabitats indicated the habitat preference of bird species. Birds are known to exploit information about the habitat patches to choose sites for feeding, shelter or reproduction to increase their realised fitness (Krebs and Kacelnik, 1991). Habitat selection is an evolved set of behaviour based on inmate responses to stimuli or learned cues. The manner in which an organism interacts, behaves in relation to the resources it needs for its existence and reproduction, competes or cooperates with others of the same or different species seeking to use the same resources and still others seeking to prey upon it, constitute an enormous field of study.

Cattle Egrets were recorded preferring the paddy fields and many others have also reported the similar behaviour (Nagarajan and Thiyagesan, 1998; Lanes and Fujioka, 1998). The use of paddy fields by Cattle Egrets also showed a strong seasonal pattern and the highest utilisation occurred during November, December and January. This was often associated with post-harvest ploughing of fields, which is comparable with results of Bredin (1983). Cattle Egrets were observed less in the open water habitat, as they were not able to catch prey in deep water (Katzir *et al.*, 1999)

Table 10.8: Microhabitat use of wetland bird species in the Kole wetlands during Wet 2 season (Newman-Keuls test) [a, b]

Variables	Microhabitat* (Mean values in increasing order) Wet-II season											
	6	**12**	**3**	**1**	**10**	**11**	**9**	**2**	**4**	**7**	**5**	**8**
Species Richness	1.67 ± 0.66	7.00 ± 6.99	7.00 ± 3.21	10.67 ± 5.36	14.67 ± 6.56	15.33 ± 9.87	11.33 ± 1.20	16.00 ± 5.03	17.00 ± 4.04	21.67 ± 7.69	22.33 ± 3.93	26.33 ± 8.20
	2	**9**	**3**	**8**	**11**	**6**	**10**	**5**	**12**	**1**	**7**	**4**
Diversity	0.90 ± 0.33	1.13 ± 0.38	1.03 ± 0.20	0.72 ± 0.39	1.15 ± 0.19	1.02 ± 0.51	1.09 ± 0.64	1.70 ± 0.56	1.19 ± 0.62	1.81 ± 0.36	1.68 ± 0.35	1.95 ± 0.13
	12	**3**	**6**	**4**	**9**	**1**	**2**	**11**	**10**	**5**	**7**	**8**
Density (Number/ha)	1923.67 ± 191.61	368.33 ± 349.48	110.67 ± 41.01	189.67 ± 105.62	291.33 ± 106.85	322.00 ± 102.31	2160.00 ± 1623.44	1280.67 ± 687.01	6698.67 ± 4707.43	3206.33 ± 1419.98	12590.67 ± 7299.53	11427.33 ± 5192.93

*1 = Floating vegetation; 2 = Bund; 3 = Bamboo pole; 4 = Tree; 5 = Electric line; 6 = Water edge; 7 = Shallow water; 8 = Mud flat; 9 = Grass; 10 = Paddy field; 11 = Open water; 12 = Harvested paddy field

a = Newman-Keuls test was carried out using log transformed data, b = The mean values given above are in original scale

Table 10.9: Microhabitat use of wetland bird species in the Kole wetlands during Migratory season (Newman-Keuls test) [a, b]

Variables	Microhabitat* (Mean values in increasing order) Migratory Season											
	6	3	1	12	9	11	2	4	7	10	5	8
Species Richness	7.33 ± 3.06	9.33 ± 2.66	16.00 ± 5.50	15.67 ± 2.66	16.67 ± 4.33	26.00 ± 6.55	28.00 ± 4.93	32.33 ± 7.79	33.67 ± 6.17	34.33 ± 4.25	38.67 ± 6.17	39.33 ± 4.97
	1	3	6	5	2	9	12	11	10	7	8	4
Diversity	1.63 ± 0.20	1.14 ± 0.22	1.15 ± 0.19	1.66 ± 0.27	1.38 ± 0.32	1.46 ± 0.12	1.55 ± 0.36	1.79 ± 0.06	1.98 ± 0.18	1.96 ± 0.15	2.33 ± 0.23	2.75 ± 0.13
	3	4	1	6	9	2	12	11	5	7	8	10
Density (Number/ha)	485.00 ± 406.73	377.67 ± 190.93	611.33 ± 215.03	747.33 ± 303.91	2945.00 ± 1346.92	5113.67 ± 2829.70	4386.00 ± 1681.50	8683.33 ± 4112.82	6630.33 ± 2243.34	21542.33 ± 10063.37	18657.67 ± 7879.72	18277.33 ± 7635.57

*1 = Floating vegetation; 2 = Bund; 3 = Bamboo pole; 4 = Tree; 5 = Electric line; 6 = Water edge; 7 = Shallow water; 8 = Mud flat; 9 = Grass; 10 = Paddy field; 11 = Open water; 12 = Harvested paddy field
a = Newman-Keuls test was carried out using log transformed data, b = The mean values given above are in original scale

Table 10.10: Microhabitat use of wetland bird species in the Kole wetlands during Non Migratory season (Newman-Keuls test)[a,b]

Variables	Microhabitat* (Mean values in increasing order) Non-migratory Season											
	6	**8**	**10**	**3**	**12**	**1**	**11**	**4**	**7**	**9**	**2**	**5**
Species Richness	3.00 ± 0.57	6.33 ± 1.76	5.00 ± 1.15	5.00 ± 0.99	11.00 ± 1.15	13.00 ± 1.73	16.67 ± 6.76	17.33 ± 4.63	16.33 ± 1.76	16.67 ± 0.88	18.33 ± 3.18	28.67 ± 1.20
	6	**3**	**11**	**10**	**12**	**9**	**2**	**1**	**8**	**7**	**5**	**4**
Diversity	0.39 ± 0.10	0.55 ± 0.10	1.22 ± 0.21	1.02 ± 0.09	1.06 ± 0.16	0.84 ± 0.42	1.65 ± 0.16	1.52 ± 0.12	1.53 ± 0.10	1.78 ± 0.38	1.92 ± 0.25	2.47 ± 0.30
	8	**4**	**3**	**10**	**6**	**2**	**12**	**9**	**7**	**5**	**1**	**11**
Density (Number/ha)	492.67 ± 428.31	66.33 ± 17.22	109.67 ± 19.34	676.67 ± 620.12	186.67 ± 94.36	468.33 ± 192.07	759.33 ± 265.52	1261.00 ± 730.05	1482.00 ± 617.87	1043.67 ± 602.56	1368.67 ± 621.49	1712.33 ± 871.95

*1 = Floating vegetation; 2 = Bund; 3 = Bamboo pole; 4 = Tree; 5 = Electric line; 6 = Water edge; 7 = Shallow water; 8 = Mud flat; 9 = Grass; 10 = Paddy field; 11 = Open water; 12 = Harvested paddy field

a = Newman-Keuls test was carried out using log transformed data, b = The mean values given above are in original scale

The observations indicated that Little Egret and Median Egret preferred the shallow water followed by the paddy fields, which is comparable with the previous studies of Nagarajan and Thiyagesan (1998). According to Kushlan (1986), the wading birds whose habitat use is closely related to the water depth were most likely to select shallow water habitats. Paddy fields are now an integral part of wetland landscapes throughout the world and such areas are often viewed as important habitat for certain species of waterbirds, particularly in the region where the availability of natural marshes had diminished (Fasola and Ruiz, 1996; Fasola *et al.*, 1996). The Cormorant and Waterfowls preferred the floating vegetation and the open water habitat. This confirms the earlier report of Esler (1992).

In the Kole wetlands, periodical fluctuations in the extent of microhabitats were common, which was very evident in the shallow water and the mud flats. The appearance of mud flats attracted large number of waders during the migratory season. Among the waders, Common Sandpiper, Wood Sandpiper and Marsh Sandpiper preferred the mud flats for foraging and such preference were reported by Kushlan (1976), Howes and Bakewell (1989), Goss-Custard and Yates (1992), Skagen and Knopf (1994) and Summer and Kalejta-Summer (1996). Other wading species were also observed in the marshy areas and they avoided feeding in the dry areas.

Overlap Index among the 11 species of birds in different microhabitats showed only fifty per cent overlap. The overlap in the habitats between different species of Egrets and Herons is a usual phenomenon and they use diverse strategies to overcome the seasonal fluctuations of water depth (Kushlan, 1986). Custer and Osborn (1978) reported that Snowy Egrets *Egretta thula* and Tricoloured Herons *Egretta tricolor* used the same sites on the North Carolina Coast. Kent (1986) reported similar habitat overlap among four species of herons. Although their habitat overlaps, they coexist in the same habitat by resource partitioning within the habitat and preying upon different species (Custer and Osborn, 1978; Kalejta, 1991).

The microhabitat utilisation pattern at Kole wetlands indicated that each species selected among the available habitats in relation to the foraging efficiency. The habitat utilisation of birds was depending on the abundance of food and habitat structure. The distribution of food is one of the most important factors influencing the selection of feeding sites by birds (Grant and Grant, 1987). Another hypothesis, which explain the foraging habitat selection pattern is that, the behavioural strategies may confine the distribution of birds in relation to features of individual habitat and landscapes.

Maintenance of habitat diversity is essential for avian diversity, which requires various water regimes and plant communities. Less appreciated habitats such as the mud flats and the shallow waters proved especially attractive to a great diversity of birds, such as waders. Schoener (1971) suggested that inter specific competition in a habitat results in emigration of birds to species niches in which they will exploit the resources more efficiently than other birds. Such specialization and shifting of niches occur when food resources become scarce (Holmes and Pitelka, 1968). Distribution of birds at Kole wetlands is in response to change in feeding conditions and difference in food availability in various microhabitats rather than to the density dependent factors mediated through competition.

Many modern studies on avian habitat selection are related to the questions of community structure, competition and other synecological concepts (Cody, 1981). It is easy to demonstrate that bird morphology is closely associated with habitat selection. Body size is one of important feature that certainly determines the habitat selection. Body size together with its physiological consequences determine diving abilities and as a consequence, the habitat selection in Cormorants (Winkler, 1983). According to Ficken and Ficken (1967), size difference may also influence dominance relationships, which in turn determine the quality of the habitat occupied. Bird population parameters such as species richness, density and diversity of birds are frequently used as indicators of habitat quality (Nilsson and Nilsson, 1978; Weller, 1978; Sampath and Krishnamoorthy, 1990, Sampath *et al.*, 1995).

Vast extent of mudflats available at Kole wetlands was the prime habitat for waders. According to Moser and Summer (1987), waders are attracted to the mudflats because these habitats support high densities of invertebrate prey. Weller (1994) also reported that the availability of the mudflats is known to contribute to the high diversity of waders. The area serves many avian species for a wide variety of purposes such as nesting, roosting and wintering ground. The study indicated that the habitat selection by wetland birds at the Kole wetlands is influenced by the prey availability and accessibility, whereas the water depth influenced the accessibility of the birds to the prey species.

❑❑❑

Chapter 11

Food and Feeding of Selected Species

Wetland birds serve an important ecosystem function of accelerating the nutrient cycling in the feeding grounds (Morales and Pacheco, 1986) and regulating fish populations (Miranda, 1995). According to Drent and Dann (1980), the availability of food is of primary importance to the birds during migration when, they store energy for winter survival and successful completion of migration. Many migrant shorebirds relay on a few key stopover sites to complete their annual migratory cycle (Myers *et al.*, 1987), which provide a unique combination of food resources and habitat necessary to support these large numbers of birds (Myers, 1984).

When animal species assemble at a particular area, they utilise individual but overlapping segments of the habitat (MacArthur, 1958). Karr (1980) stated that resource utilization pattern and the community organization are determined by a combination of variables. The availability and preference of the food organisms, feeding behaviour and utilization of the habitat for feeding and reproduction differ in co-existing animal species. Periods of rapid change in prey populations offer important opportunities to study the relationship between predators and their prey. In particular, such studies can yield information on the functional response of predators to a changing prey environment and the role of prey dynamics in the regulation of predator population size.

Biologists need to better understand the factors influencing the use of wetlands by migratory birds including effects of invertebrate distribution and abundance on habitat selection by non-breeding birds (Combs and Fredrickson, 1996). Presence and abundance of aquatic invertebrates may help to explain why diverse and abundance of migratory birds are more attracted to moist soil impoundment than the flooded row crops in some regions (Fredrickson and Taylor, 1982). Shallow water wetlands constitute important foraging habitats for wading birds (Ardeidae) (Kushlan, 1978; Ramo and Busato, 1993). Many of the studies on foraging behaviour of wetland birds have been designed to test the predictions of optimal foraging theory. This theory proposes that an animal should forage in an energy efficient manner, which maximises its survival and therefore its genetic contribution to subsequent generations. Optimal foraging can be attained through choice of prey, selection of feeding patch, allocation of time and optimal pattern and speed of movement. This chapter described the food and feeding habits of selected wetland species and the food availability in the form of benthic and macro fauna in Kole wetlands.

METHODS

Estimation of prey size

Direct observation method was used for studying the food and feeding patterns of five selected species of birds (Altman, 1974). Direct observation of foraging is an effective technique for studying the diet of wetland birds. This method does not require the capture or killing of birds and is unaffected by different digestion rates of different prey species. However, it depends upon the ability of the observer to correctly identify and estimate the size of the prey items consumed by the bird (Cezilly and Wallace, 1988). The birds quickly accepted the presence of the observer, often feeding within 200 m from the observer. Foraging observations for each species was recorded in all the three seasons. During the observations, it could reliably determine if foraging pecks were successful even for small sized prey items. Observations were made from 0600 hour to 1730 hour and the information was collected from all the areas, opportunistically. Sample size varied from 231 to

356 counts per species. Prey length was estimated by comparison of the prey length with various dimensions of the bird head and bill morphology (Recher and Recher, 1972). The size of prey was determined 0.25x, 0.5x, 1x, 2x, 3x, 4x etc. Bill length was collected from the literature (Ali and Ripley, 1983) and also through measuring the dead specimens of birds from the Kole wetlands. Whenever a successful feeding attempt was observed, prey length was assessed. Similarly, when the bird moved away from the feeding area, the water depth, where the bird was foraging was measured.

Estimation of food availability

Estimation of benthic fauna : The mud samples were collected from different microhabitats *viz.* shallow water, mud flats and paddy fields during the migratory seasons by Naturalist's dredge method (size 32.14 cm, depth 32 cm) in the intensive study areas. The dredge was towed at a minimum speed and for a distance of 30 cm. The mud samples were collected from the plots identified for censusing the birds. From each plot 5 samples were collected and the collected mud was sieved through 0.5 mm sieve and the fauna were filtered and preserved in 5 per cent formaldehyde (Strin, 1981). The benthic animals were identified up to Class by using Ward and Whipple (1945) and Hyman (1967).

Estimation of macro fauna : Macro fauna was collected from the shallow water, mudflats and paddy fields using quadrat method. The size of the quadrat was 1 m X 1 m at the intensive study areas and the frequencies of each animal category for each microhabitat were recorded. Samples were collected from the plots uniformly representing all the areas. The collection was done only during migratory season. In each month, five samples (quadrat) were collected from each identified plots in the intensive study areas. Prey items were identified up to broad taxonomic categories such as Crabs and Snails (Hafner *et al.*, 1986). The collected organisms were preserved in 20 per cent formaldehyde. Sampling was done in all habitats in each month and the abundance of each category for each habitat was recorded.

Analysis of variance (ANOVA) (Gomez and Gomez, 1984) was performed to examine the variation of abundance in benthic and macro fauna over the years in different habitats. The analysis was carried out with log-transformed data.

Prey size estimation

Prey size of six species of birds namely Little Cormorant, Indian Pond- Heron, Little Egret, Large Egret and Whiskered Tern were studied. The details of each species are given below.

Little Cormorant (*Phalacrocorax niger*)

During the southwest monsoon, when the whole region was inundated the Little Cormorants were seen in most of the area diving for fishes. However, during the summer, when the water was restricted to the canals, all the birds congregated near the canals. The main method of feeding of the species was by diving and catching the fishes under water. The major food of Little Cormorant is fish and to some extent tadpoles and crustaceans. In the present study, the length of the fishes caught by Little Cormorant ranged from 6 to 55 cm and the highest preference was recorded for fishes having a length of 12 cm (Fig. 11.1). The highest number of foraging attempts was recorded in the 101 to 150 cm water depth (Fig. 11.2). During the monsoon, the bird was diving in the deep waters for fishing while in summer, it was depending on canals for food.

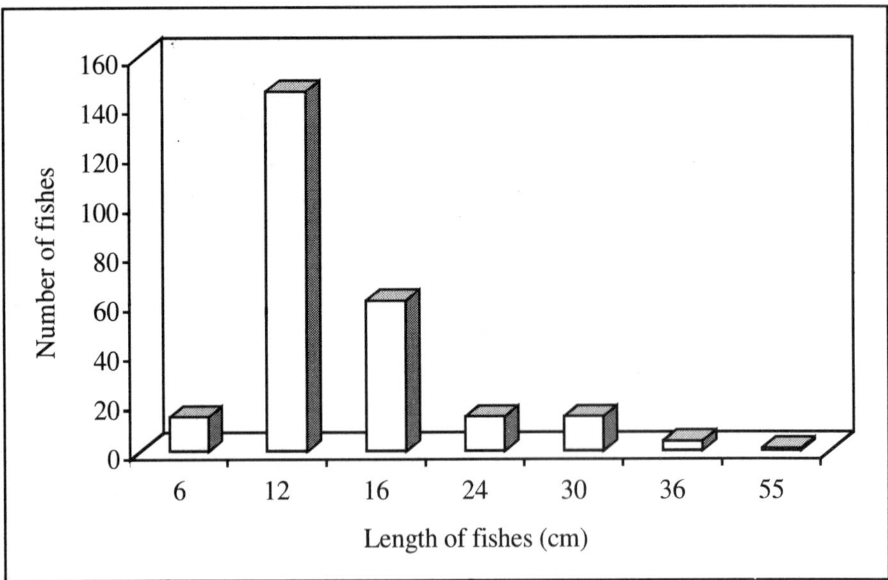

Fig. 11.1: Prey length preference of Little Cormorant in the Kole wetlands (n = 254)

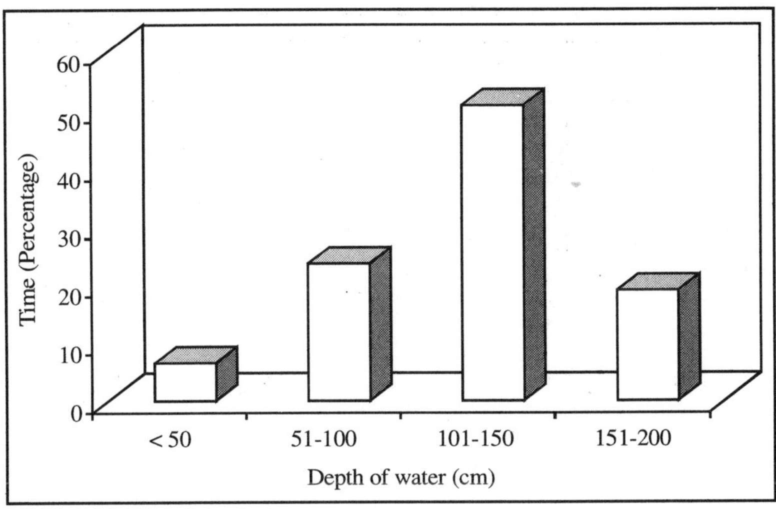

Fig. 11.2: Preferred water depth of foraging by Little Cormorant in Kole wetlands (n = 415 hours)

Indian Pond-Heron (*Ardeola grayii*)

This was another common bird in the Kole wetlands. They were feeding singularly in the shallow waters and its main food items were fish, frog, crustaceans and insects. Pond Heron generally waited with a poised neck at the edge of water and as soon as a prey was visible, it striked with its beak and captured the prey. The length of fishes varied from 2 to 40 cm and the preferred prey was fishes of 7 cm (Fig. 11.3). The Pond Heron preferred water, with a depth of less than 10 cm. Next preferred category of water depth was 11 to 20 cm (Fig. 11.4).

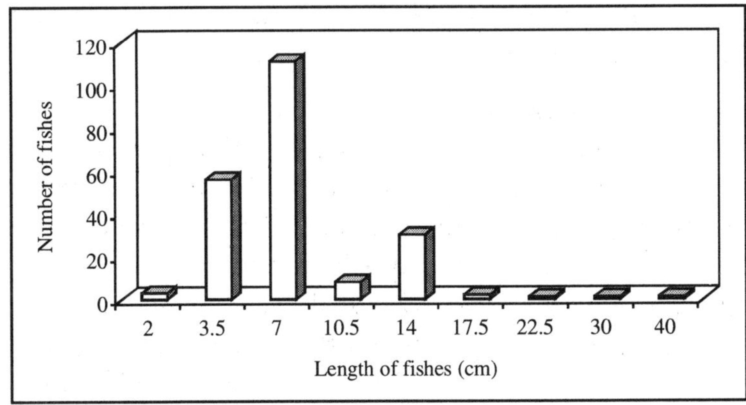

Fig. 11.3: Prey length preference of Indian Pond-Heron in the Kole wetlands (n = 213)

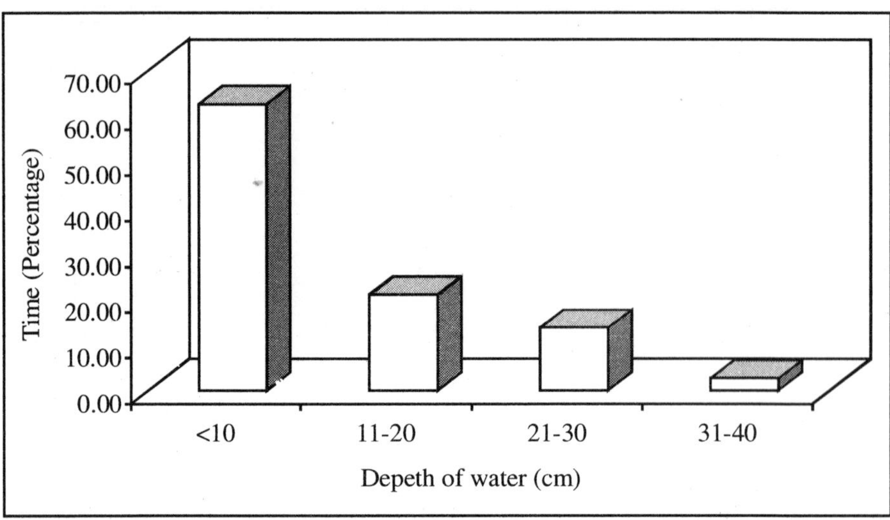

Fig. 11.4: Preferred water depth of foraging by Indian Pond-Heron in the Kole wetlands (n = 300 hours)

Little Egret (*Egretta garzetta*)

Little Egret showed varied feeding patterns and slow wading was the most frequent style of feeding behaviour of this species. It preferred shallow waters, marshes and paddy fields. They fed in small flocks or large flocks numbering about 20 to 1000 individuals. Main food items consisted of fish, frog and crustaceans. Preferred length of fish ranged from 5 to 60 cm with highest preference of fish size was observed 20 cm (Fig. 11.5). Preferred water depth of feeding varied from < 10 cm to 50 cm and the highest incidence of foraging was observed in < 10 cm water depth (Fig. 11.6).

Two types of foraging were recorded in the species. Little Egret usually adopted food string method while walking, shaking their foot with each step and striking, as prey was disturbed. Another method of foraging was active pursuit, which included galloping chases, flapping of wings and many quick turns. Strikes were frequent, missing more often than not. After strike, the birds usually appeared to relocate prey and then continued the pursuit. Little Egret usually stood and waited, when they were part of large feeding flocks. The bill vibrating behaviour was not common and very rarely sighted during the study.

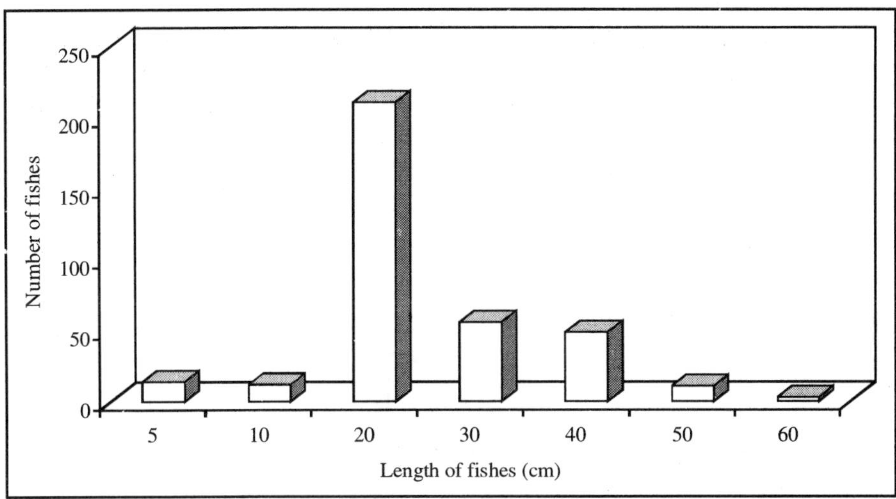

Fig. 11.5: Prey length preference of Little Egret in the Kole wetlands (n = 356)

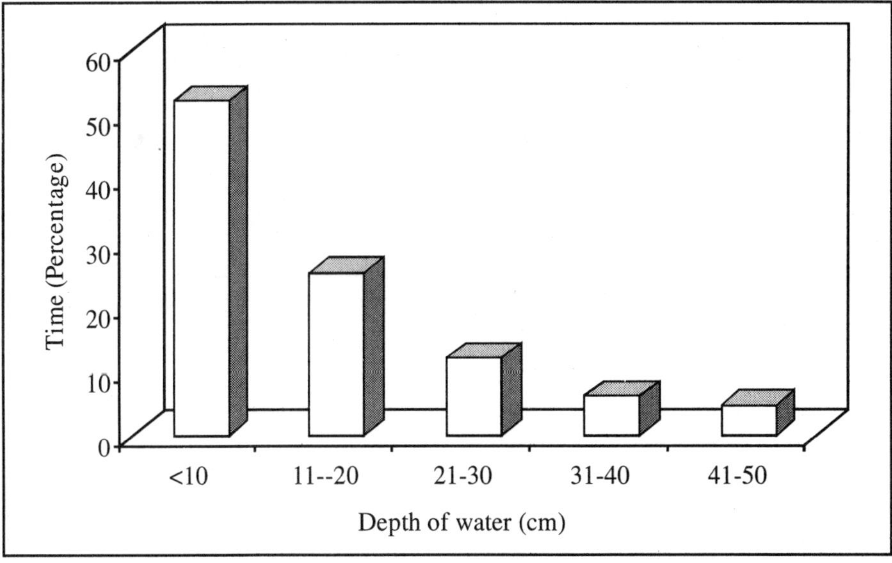

Fig. 11.6: Preferred water depth of foraging by Little Egret in the Kole wetlands (n = 300 hours)

Large Egret (*Casmerodius albus*)

Large Egret mainly fed on fishes and crustaceans and different foraging was observed in the species. They were observed to go for slow wading and upright stand and wait. Slow wading is one of the primary methods of hunting. Sometimes the bird struck the prey while moving and often swallowed the prey while wading. Another hunting method of the species was upright stand and wait behaviour. Large Egret frequently used to stand, wait and feeding the prey item in shallow water bodies. Preferred prey length of Large Egret ranged from 5 cm to 60 cm with highest preference for fishes of 30 cm length (Fig. 11.7). The preferred water depth, for foraging was from 10 cm to 50 cm (Fig. 11.8).

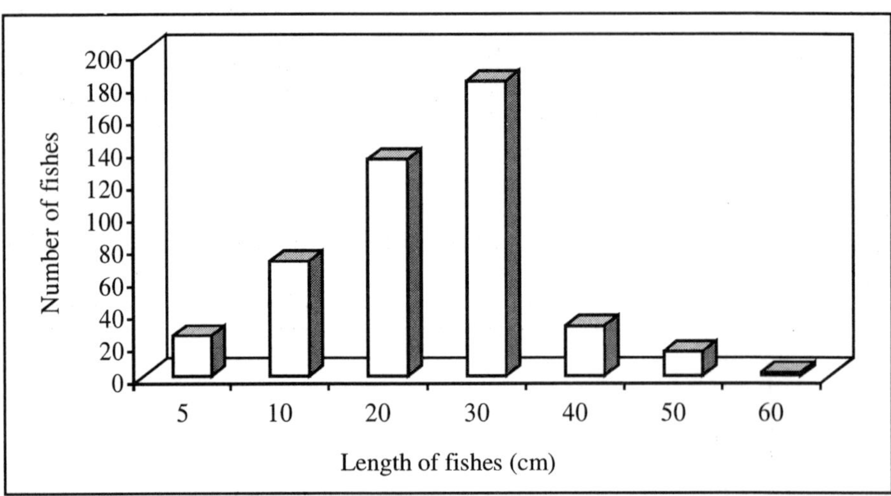

Fig. 11.7: Prey length preference of Large Egret in the Kole wetlands (n = 460)

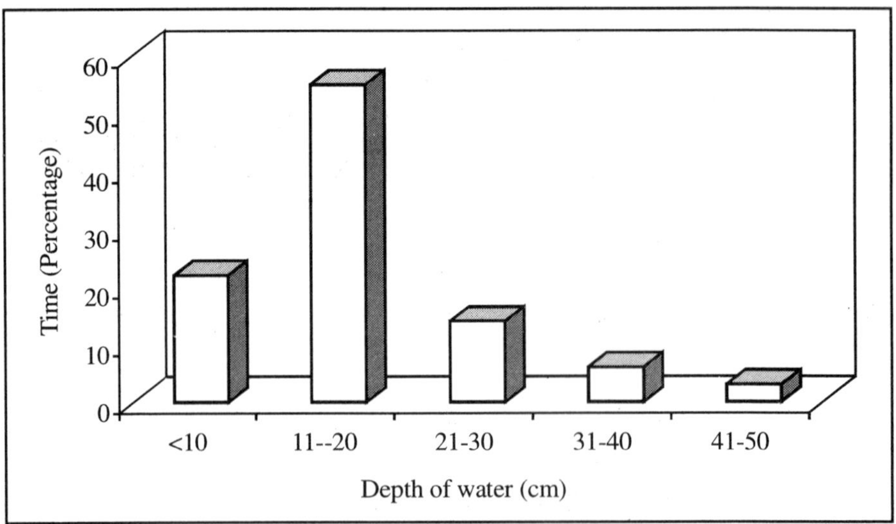

Fig. 11.8: Preferred water depth of foraging by Large Egret in the Kole wetlands (n = 200 hours)

Whiskered Tern (*Chlidonias hybridus*)

This migratory species from the lower reaches of the Himalayas were mainly concentrating in the canals for feeding. During summer, these birds follow a peculiar formation to hunt the fishes in the canals. The birds in a flock search for the fishes while flying, starting from one end of the canal to the other. Then they come back to the starting point again and do this combing operation again. The feeding was continued until darkness. After spotting a fish, the bird dived deep into the water, catching it from under water and fed it while in flight. Large flocks of Whiskered Terns (> 500 individuals) were sighted above the canals when the water is shallow. The preferred prey size of the Whiskered Tern ranged from 1.2 to 5 cm with highest of 2.5 cm length of fishes (Fig. 11.9). Preferred foraging water depth varied from < 50 cm to 200 cm and high preference was recorded for foraging in 101 to 150 cm of depth (Fig. 11.10).

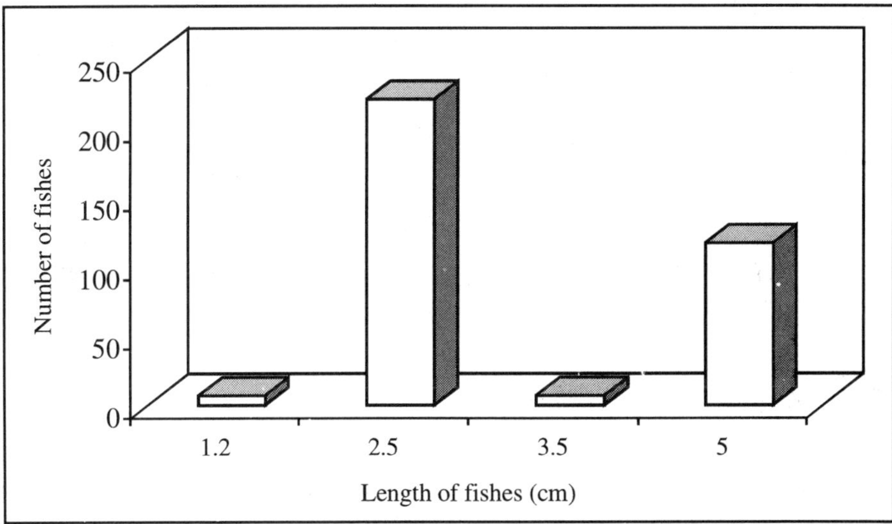

Fig. 11.9: Prey length preference of Whiskered Tern in the Kole wetlands (n = 352)

Black-shouldered Kite (*Elanus caeruleus*)

Among the raptors, Black-shouldered Kite was an occasional visitor to the Kole wetlands. The usual prey consists of locust, grasshopper, crickets, lizards, field rats, mice and young birds (Ali and Ripley, 1983). Lamba (1970) had reported Black-winged Kite catching a wounded Green Pigeon *Treron phoenicoptera* during flight. A Black-shouldered Kite was observed capturing a Wood Sandpiper from the paddy field on 6th January 2000. After capturing the prey, the Kite landed on a nearby bund. The Wood Sandpiper was alive and the Kite tried to kill the prey. The Kite removed the feathers from the wings and consumed the flesh and bones and took 35 minutes for completely feeding the prey. The Wood Sandpiper was caught from a flock of birds numbering around 50.

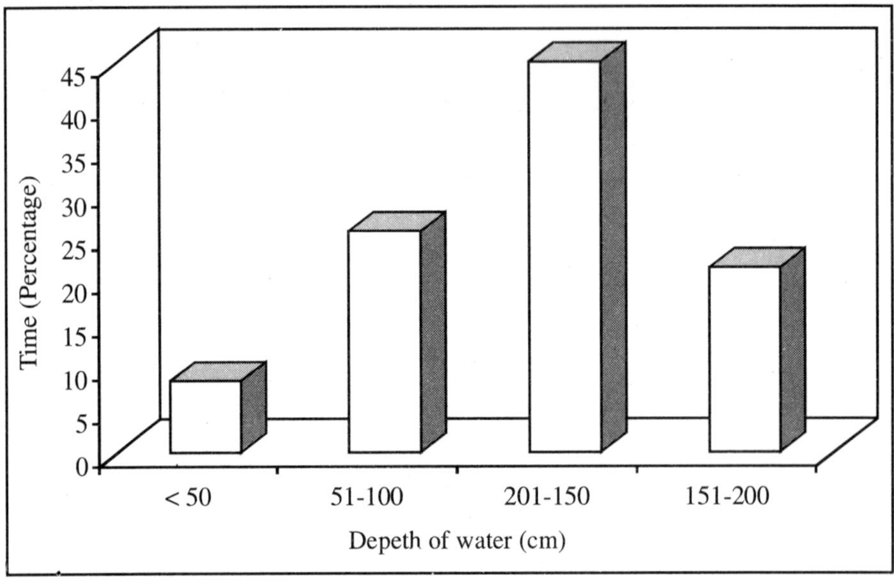

Fig. 11.10: Preferred depth of foraging by Whiskered Tern in the Kole wetlands (n = 400 hours)

Estimation of food availability

Benthic fauna

Four groups of benthic fauna were recorded from the mud samples namely Insecta, Polycheata, Gastropoda and Crustacea.

Mud flat

Four groups of benthic fauna were recorded from the mud flats. The mean monthly abundance showed that Polycheates were the dominant in all the months followed by Insecta (Table. 11.1).

Shallow Water

This microhabitat harboured four groups of benthic fauna. Highest occurrence was recorded for Insecta (325), Gastropoda (308),

Ploycheata (293) in the month of October and Crustacea (192) in February (Table 11.2).

Table 11.1: Monthly abundance of benthic fauna in the mud flats during the three migratory seasons (September to March) n=21

Name of the group	S	O	N	D	J	F	M
Insecta	233	320	440	401	243	182	323
Polycheta	342	539	363	476	551	381	257
Gastropoda	195	69	195	198	199	227	258
Crustacea	147	88	201	32	141	142	73
Unidentified	49	25	22	20	34	19	25

Table 11.2: Monthly abundance of benthic fauna in the shallow water during the three migratory seasons (September to March) n=21

Name of the group	S	O	N	D	J	F	M
Insecta	314	325	261	207	166	116	216
Polycheta	280	293	240	287	273	243	287
Gastropoda	174	308	235	239	274	219	203
Crustacea	167	172	56	89	113	192	174
Unidentified	13	3	23	17	11	7	21

Paddy field

Paddy fields were another important microhabitat in the Kole wetlands. Four groups of benthic fauna were recorded and Polycheates were recorded highest followed by Insecta (Table 11.3).

Table 11.3: Monthly abundance of benthic fauna in the Paddy fields during the three migratory seasons (September to March) n=21

Name of the group	S	O	N	D	J	F	M
Insecta	241	230	297	428	460	257	170
Polycheta	353	262	360	384	481	461	323
Gastropoda	312	171	171	213	178	321	395
Crustacea	167	201	184	221	238	232	152
Unidentified	10	5	8	11	18	12	6

Macro fauna

Two groups of macro fauna were recorded from the mud flats, shallow water and paddy fields, namely the Snails and Crabs. Snails were highest in number in all the three microhabitats. (Table 11.4, 11.5 and 11.6.).

Table 11.4: Monthly abundance of macro fauna in the mud flats during the three migratory seasons (September to March) n=21

Name of the group	S	O	N	D	J	F	M
Snail	367	593	543	757	320	401	374
Crab	86	54	31	37	71	56	71
Total	**453**	**647**	**574**	**794**	**391**	**457**	**445**

Table 11.5: Monthly abundance of macro fauna in the shallow water during the three migratory seasons (September to March) n=21

Name of the group	S	O	N	D	J	F	M
Snail	360	493	315	390	368	435	357
Crab	15	32	12	36	16	22	73
Total	**375**	**525**	**327**	**426**	**384**	**457**	**430**

Table 11.6: Monthly abundance of macro fauna in paddy fields during the three migratory seasons (September to March) n=21

Name of the group	S	O	N	D	J	F	M
Snail	391	329	303	437	473	473	352
Crab	10	11	11	28	18	17	15
Total	**401**	**340**	**314**	**465**	**491**	**490**	**467**

Influence of the microhabitat and year on the abundance of benthic fauna

Two way analysis of variance (ANOVA) was performed on abundance of benthic fauna to detect significant difference between the habitats and the year. There was significant difference between the habitat and year in the abundance of Insecta, Polycheata, Gastropoda and Crustacea. Interaction between the habitat and year indicated that the abundance of Insecta in different habitats varied significantly over the years. (Table 11.7 to 11.10).

Table 11.7: Analysis of Variance for benthic fauna - Insecta

Source	DF	SS	MSS	F
Habitat	2	0.537	0.268	3.316*
Year	2	0.737	0.369	4.553**
Habitat * Year	4	1.570	0.393	4.849**
Error	36	2.915	0.080	
Total	**44**	**5.759**		

** = <0.01; * = 0.05

Table 11.8: Analysis of Variance for benthic fauna - Polycheta

Source	DF	SS	MSS	F
Habitat	2	1.365	0.682	10.633***
Year	2	1.612	0.806	12.559***
Habitat * Year	4	0.348	0.087	1.356^{NS}
Error	36	2.310	0.064	
Total	**44**	**5.635**		

*** = <0.001; NS = Not significant

Table 11.9: Analysis of Variance for benthic fauna - Gastropoda

Source	DF	SS	MSS	F
Habitat	2	0.642	0.321	3.731*
Year	2	0.410	0.205	2.386*
Habitat * Year	4	0.591	0.148	1.718^{NS}
Error	36	3.097	0.086	
Total	**44**	**4.740**		

* = <0.05; NS = Not significant

Table 11.10: Analysis of Variance for benthic fauna - Crustacea

Source	DF	SS	MSS	F
Habitat	2	30.416	15.208	5.773*
Year	2	2.275	1.137	0.432^{NS}
Habitat * Year	4	9.141	2.285	0.867^{NS}
Error	36	94.844	2.635	
Total	**44**	**136.676**		

* = <0.05; NS = Not significant

Influence of the microhabitat and year on abundance of macro fauna

Two way analysis of variance (ANOVA) was performed on abundance of macro fauna to test the significant difference between the habitats and the year. There was no significant difference between the habitats and year in the abundance of Snail. Macro fauna of Crab showed that significant difference between the habitats. However there was no significant difference between the years and interaction between the year and the habitat (Tables 11.11 and 11.12).

Table 11.11: Analysis of Variance for macro fauna - Snail

Source	DF	SS	MSS	F
Habitat	2	0.150	0.075	0.500[NS]
Year	2	0.188	0.094	0.626[NS]
Habitat * Year	4	0.435	0.109	0.725[NS]
Error	36	5.401	0.150	
Total	**44**	**6.173**		

NS = Not significant

Table 11.12: Analysis of Variance for macro fauna - Crab

Source	DF	SS	MSS	F
Habitat	2	11.432	5.716	10.690***
Year	2	2.189	1.095	2.047*
Habitat * Year	4	0.641	0.160	0.300[NS]
Error	36	19.250	0.535	
Total	**44**	**33.512**		

*** = 0.001; * = <0.05; NS = Not significant

Fish

Thirteen species of fishes were identified from the Kole wetlands. They were *Xenetodon cancila, Garra mullya, Chanda thomassi, Etroplus suratensis, Etroplus maculates, Chela clupoides, Macropodus cupanus, Mastacembellus guntheri, Rasbora daniconius, Hyporamphus xanthopterus, Puntius filamentosus, Mystus gulio* and *Puntis pinnauratus*.

Population fluctuation of certain migratory species in relation to benthic fauna

Common Sandpiper (*Actitis hypoleucos*)

The Common Sandpiper was recorded in the Kole wetlands from September to March and the population size differed annually. Highest number of birds was recorded in the second year (1999-2000) (Fig. 11.11). The abundance of Common Sandpiper and benthic fauna was negatively correlated with Crustacea (r = -0.68, P<0.05, n=7) in the mudflats. However, there was no significant correlation with Insecta, Polycheata and Gastropoda. This was correlated with Polycheata (r = 0.57, P<0.10, n=7) in shallow water and not significant with Insecta, Gastropoda and Crustacea. Insecta, Polycheata, Gastropoda and Crustacea were not significantly correlated with the abundance of Common Sandpiper in the Paddy fields.

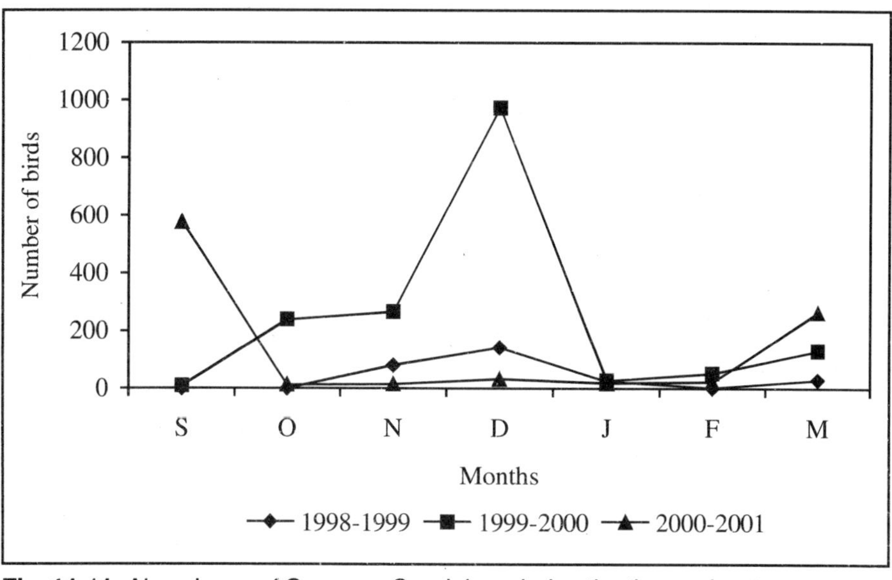

Fig. 11.11: Abundance of Common Sandpiper during the three migratory seasons in the Kole wetlands

Curlew Sandpiper (*Calidris ferruginea*)

Curlew Sandpiper was recorded only in two seasons (1999-2000 and 2000-2001). Highest abundance was recorded in December (1999) and January (2001) and the lowest was in March (2001) (Fig. 11.12). The population size of Curlew Sandpiper showed significant correlation with Insecta (r = 0.72, P<0.02, n=7) in Mud flats and there was no significant correlation with Polycheata, Gastropoda and Crustacea. There was significant correlation with Crustacea (r = -0.65, P<0.05, n=7) in Shallow water, but no significance was observed with Insecta, Ploycheata and Gastropoda in Shallow water and Paddy fields.

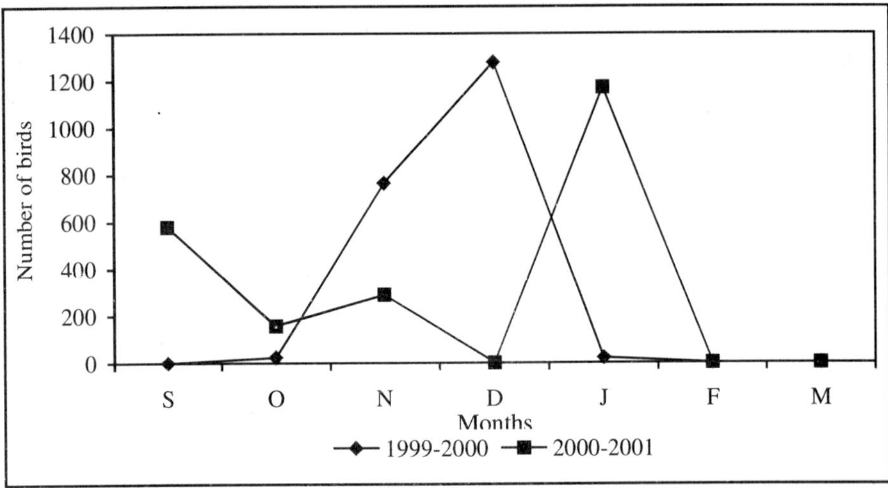

Fig. 11.12: Abundance of Curlew Sandpiper during the three migratory seasons in the Kole wetlands

Little Ringed Plover (*Charadrius dubius*)

There were monthly variations in the abundance of Little Ringed Plover and two peaks in the population in November and January during 2000-2001 (Fig. 11.13). A significant correlation was found between the population size of Little Ringed Plover and abundance of Insecta (r = 0.70, P<0.02, n=7) and Crustacea (r = -0.58, P<0.10, n=7) in Mud flats, however there was no correlation was found with polycheata and Gastropoda. It was significantly correlated with Polycheata (r = -0.59, P<0.10, n=7), Crustacea (r = -0.77, P<0.02, n=7) in the Shallow water and it was not correlated with Insecta and Gastropoda. No significant correlation was found with Insecta, Polycheata, Gastropoda and Crustacea in the Paddy fields.

Eurasian Curlew (*Numenius arquata*)

This species arrived in the study area during September in the second (1999-2000) and third year (2000-2001) mainly attracted to the shallow water habitat. The highest abundance was recorded in November (1998-1999) and in December (1999-2000) (Fig. 11.14). The population of Curlew was significantly correlated with Insecta (r = 0.69, P<0.05, n=7) in Mud flats and it was not correlated with Polycheata, Gastropoda and Crustacea. There was significant correlation with Crustacea (r = -0.76, P<0.02, n=7) in the shallow water and correlation was not significant with other groups like Insecta, Polycheata and Gastropoda. The population of Curlew was significantly correlated with Gastropoda (r = -0.57, P<0.10, n=7) in the Paddy fields, and there was no correlation with Insecta, Polycheata and Crustacea.

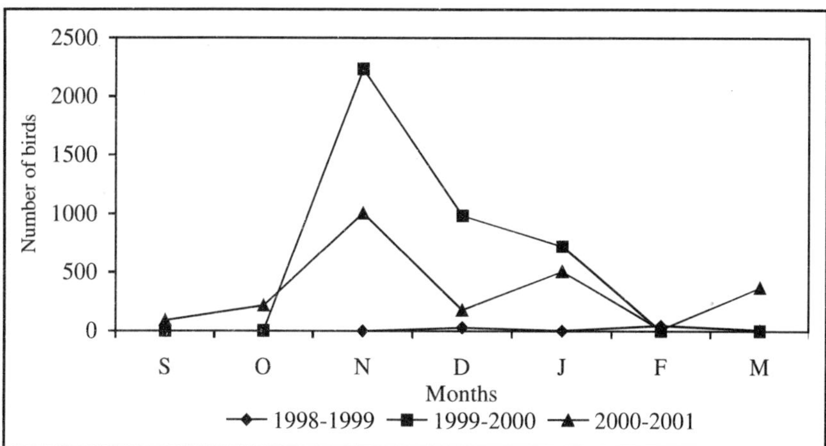

Fig. 11.13: Abundance of Little Ringed Plover during the three migratory seasons in the Kole wetlands

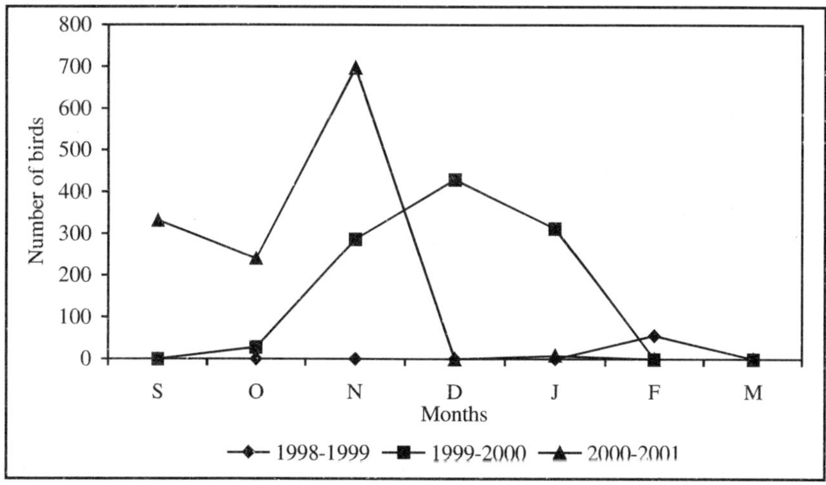

Fig. 11.14: Abundance of Eurasian Curlew during the three migratory seasons in the Kole wetlands

Little Stint (*Calidris minuta*)

The emergence of mud flats attracted this species into the Kole wetlands. The highest number of birds was observed in November (1999-2000) and September (2000-2001) (Fig. 11.15). Two peaks of population were observed during the year 2001. The first one was in September and the second in November. The change of the preferred microhabitat into non-preferred one showed the decline in number of birds, which happened in February and March throughout the study period. Population size of Little Stint was found significantly correlated with Crustacea (r = 0.59, P<0.10, n=7) in the Mud flats. It was also correlated with Crustacea (r = -0.54, P<0.10, n=7) in the Shallow water and there was no significant correlation with other groups in other habitats.

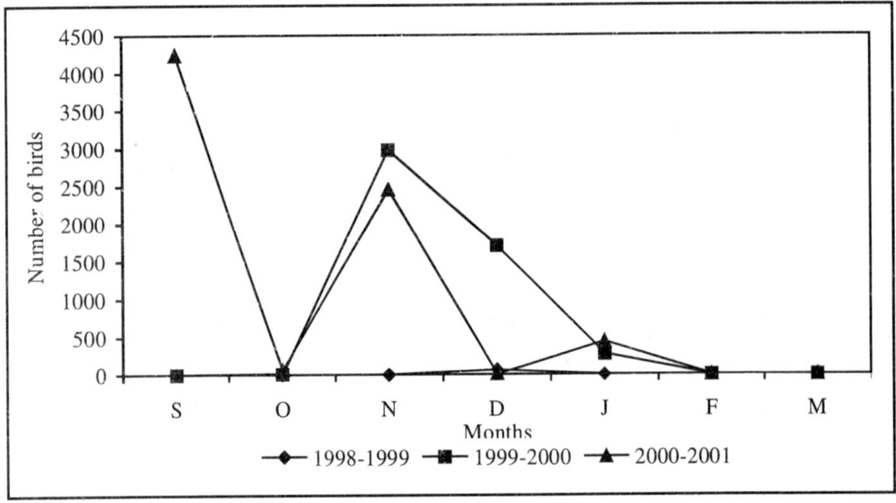

Fig. 11.15: Abundance of Little Stint during the three migratory seasons in the Kole wetlands

Wood Sandpiper (*Tringa glareola*)

This is the most common wader species in the Kole wetlands and they arrived much earlier than the other species. The peak in the population was observed in December (1999-2000) and in November (2000-2001) (Fig. 11.16). Abundance of Wood Sandpiper was found significantly correlated with Insecta (r = 70, P<0.02, n=7) in Mud flats and was not correlated with Polycheata, Gastropoda and Crustacea. Significant correlation was observed with Crustacea (r = -0.82, P<0.01, n= 7) in the Shallow waters and no correlation with Insecta, Polycheata and Gastropoda. Significant correlation was found between population of Wood Sandpiper and Insecta ($r = 0.61$, $P<0.05$, $n=7$), Gastropoda ($r = -0.51$, $P<0.10$, $n=7$) in the Paddy fields, whereas it was not significant with Ploycheata and Crustacea.

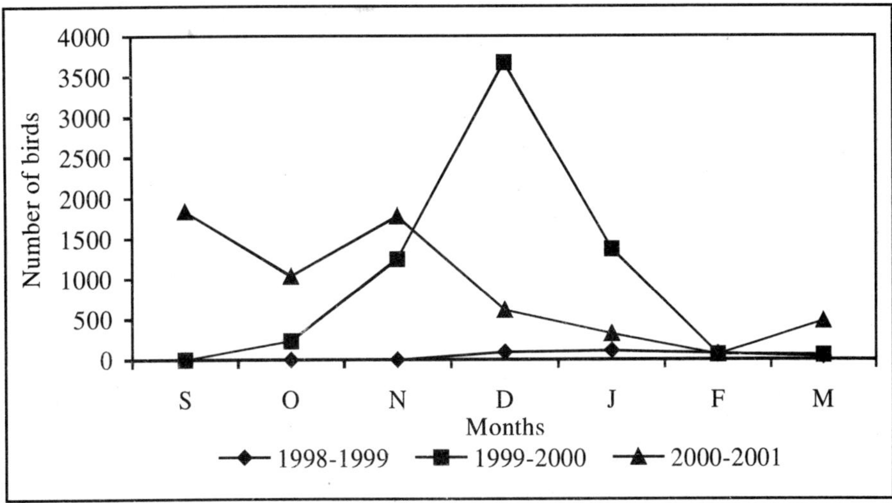

Fig. 11.16: Abundance of Wood Sandpiper during the three migratory seasons in the Kole wetlands

Common Redshank (*Tringa totanus*)

This species was recorded in the Kole wetlands from September to March in all the year. The population size fluctuated every month and a decline was observed in February and March (Fig. 11.17). Two peaks in population were observed, one was in November and the other in January during 1999-2000. The population size of Common Red Shank was significantly correlated with the population of Insecta ($r = 0.58$, $P<0.10$, $n=7$), Crustacea ($r = 0.68$, $P<0.05$, $n = 7$) in Mud flats, whereas no significant correlation was observed with Polycheata and Gastropoda. Significant correlation was observed with Polycheata ($r = -0.63$, $P<0.05$, $n = 7$) and Crustacea ($r = -0.74$, $P<0.02$, $n = 7$) in Shallow water, but no significant correlation found with Insecta and Gastropoda.

Common Greenshank (*Tringa nebularia*)

This species was observed in the Kole wetlands from September to March and utilised shallow waters and mud flats. The peaks were in November and February during 1999-2000 (Fig. 11.18). Significant correlation was found with the abundance of Common Green Shank and Gastropoda (r = 0.60, P<0.05, n=7) in Mud flats and it was not significantly correlated with Insecta, Polycheata and Crustacea. Similarly, significant correlation was found with Insecta (r = 0.55, P<0.10, n = 7) in the Shallow water and there was no significant correlation with other groups in the Shallow water and Paddy fields.

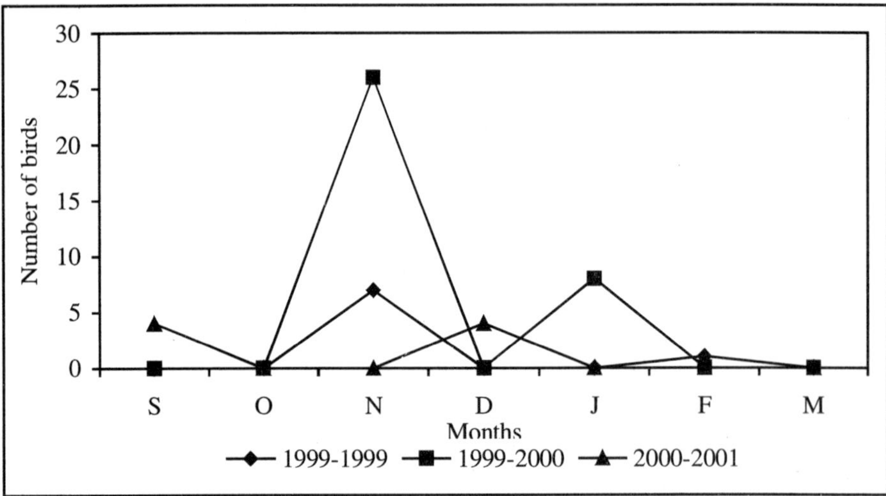

Fig. 11.17: Abundance of Common Redshank during the three migratory seasons in the Kole wetlands

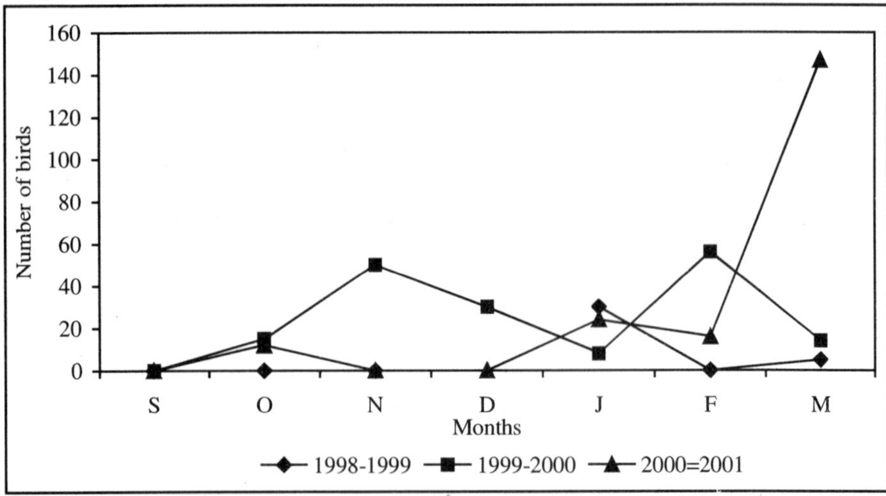

Fig. 11.18: Abundance of Common Greenshank during the three migratory seasons in the Kole wetlands

Green Sandpiper (*Tringa ochropus*)

Green Sandpiper was recorded from September to March during the second (1999-2000) and third year (2000-2001), but no bird was observed in the first year. The peak population was observed in September (2000) (Fig. 11.19). The abundance of Green Sandpiper was significantly correlated with Gastropoda (r = -0.92, P<0.01, n=7) in the Mud flats. Significant correlation was observed with Insecta (r = 0.55, P<0.10, n=7) and Gastropoda (r = 0.74, P<0.02, n=7) in the Shallow waters. In the Paddy fields this was significantly correlated with Polycheata (r = -0.66, P<0.05, n=7), Gastropoda (r = -0.53, P<0.10, n=7) and Crustacea (r = -0.54, P<0.10, n=7).

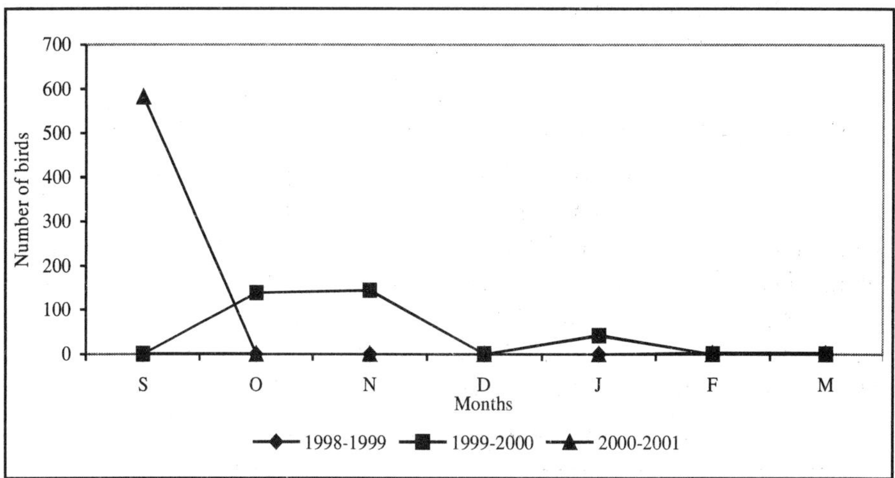

Fig. 11.19: Abundance of Green Sandpiper during the three migratory seasons in the Kole wetlands

Small Pratincole (*Glareola lactea*)

Small Pratincole was observed throughout the study period, but highest population was recorded during January 1999-2000 (Fig. 11.20). Significant positive correlation was observed with Polycheata (r = 0.72, P<0.02, n=7) in the Mud flats. The correlation with Crustacea (r = - 0.58, P<0.10, n=7) was also significant in the Shallow waters. There was significant correlation between abundance of Small Pratincole with Insecta (r = 0.72, P<0.01, n=7), Polycheata (r = 0.6, P<0.05, n=7), Crustacea (r = 0.80, P<0.01, n=7) and Gastropoda (r = - 0.55, P<0.10, n=7) in the Paddy fields.

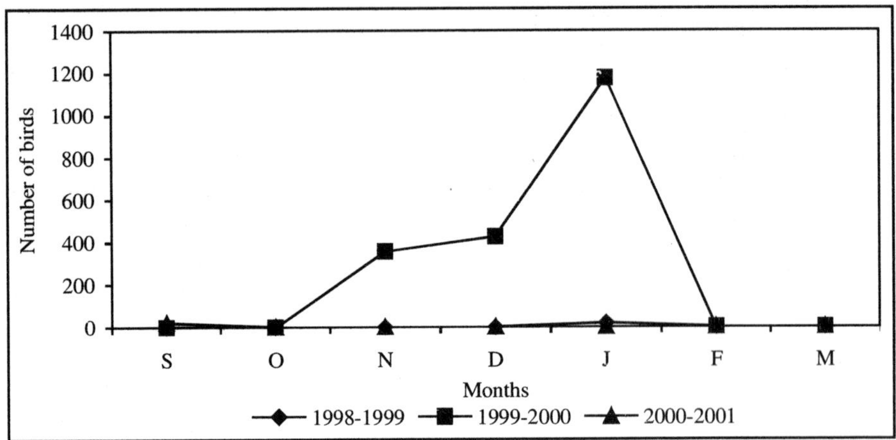

Fig. 11.20: Abundance of Small Pratincole during the three migratory seasons in the Kole wetlands

Black-winged Stilt (*Himantopus himantopus*)

They were found in the Kole wetlands from September to March in all the migratory seasons. The peak population was recorded in December (1999-2000) and March (2000-2001) (Fig. 11.21). Abundance of Black-winged Stilt was found significantly correlated with Insecta ($r = 0.59$, $P<0.10$, $n=7$) and Crustacea ($r = - 0.67$, $P<0.05$, $n=7$) in the Mud flats. No significant correlation was found with other groups in the Shallow water and Paddy fields.

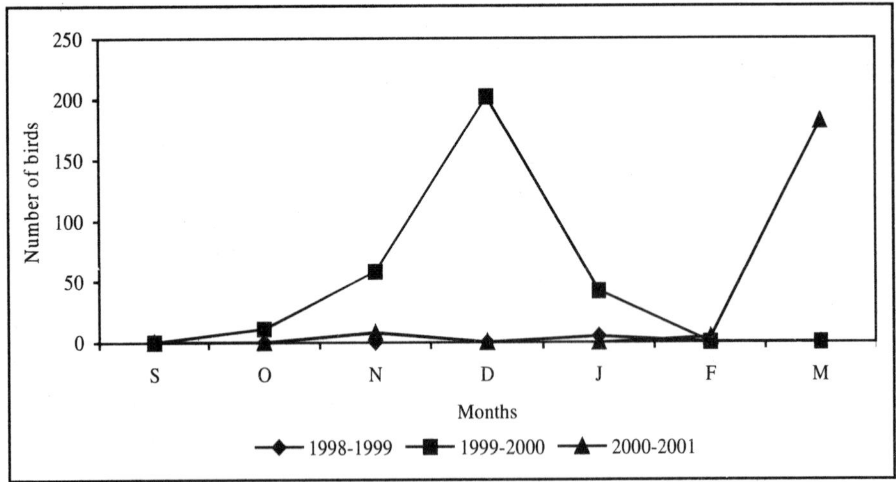

Fig. 11.21: Abundance of Black-winged Stilt during the three migratory seasons in the Kole wetlands

Marsh Sandpiper (*Tringa stagnatilis*)

Marsh Sandpiper arrived in the Kole wetlands after the appearance of mud flats and shallow water and the highest peak was in November (2000) decreasing gradually from January in all years (Fig. 11.22). Significant correlation was observed between the abundance of birds with Insecta and Crustacea (r = 0.64, P<0.05, n=7) in Mud flats, no significant correlation existed between the abundance of Polycheata and Gastropoda. Similarly, significant correlation was found with Polycheata (r = -0.65, P<0.05, n=7), Crustacea (r = -0.71, P<0.02, n=7) in the Shallow water.

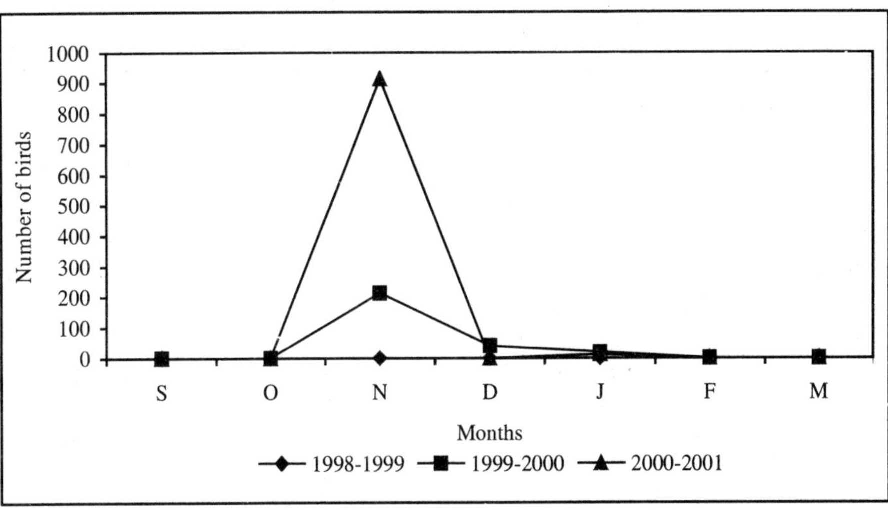

Fig. 11.22: Abundance of Marsh Sandpiper during the three migratory seasons in the Kole wetlands

Lesser Sand Plover (*Charadrius mongolus*)

Lesser Sand Plover arrived in September after the emergence of mud flats and shallow water. The peak population was observed in September and November during the year 2000-2001. The highest population was recorded in September (2000-2001) and the species was not recorded in February and March (Fig. 11.23). Abundance of Lesser Sand Plover was found significantly correlated with Insecta (r = 0.63, P<0.05, n=7) and Crustacea (r = 0.66, P<0.05, n=7) in Mud flats. Similarly, significant correlation was observed in the abundance of Lesser Sand Plover with Polycheata (r = -0.65, P<0.05, n=7) and Crustacea (r = -0.69, P<0.05, n=7) in Shallow water. No significant correlation was found between benthic fauna in the Paddy fields.

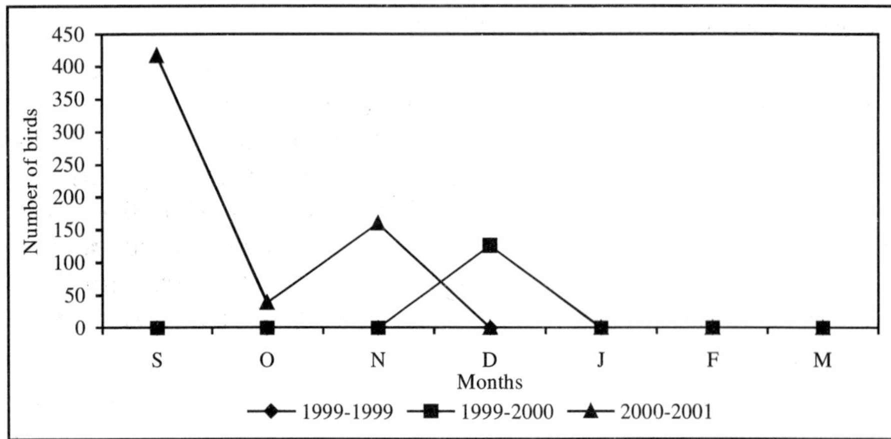

Fig. 11.23: Abundance of Lesser Sand Plover during the three migratory seasons in the Kole wetlands

Food and feeding behavior studies showed that sufficient prey is available in the Kole wetlands. Availability of small fishes will boost the food resources of piscivorous birds. In the Kole wetlands the food selection of wetland species depended on the differences in the feeding technique, body size and bill morphology. Prey length of Little Cormorant varied from 6 cm to 54 cm, which is comparable with earlier studies of Mukherjee (1969). Little Cormorant is usually described as an opportunistic predator, which means that the composition of its diet is determined by the availability of fish species (Cooper, 1985).

Indian Pond-Heron fed fishes having a length from 2 cm to 37.5 cm. Sodhi (1986) has showed that food items of Pond Heron and Little Egret varied from 2 mm to 111 mm, which is comparable with the present observation. Pond Heron is predominantly a solo feeder and its plumage is cryptic. The cryptic plumage for a wading bird is an adaptation for solo feeding (Kushlan, 1987). Basic technique of the species is waiting and Pond Heron waits for the prey to come in the strike zone. The chances of catching prey decreases, if two or more birds are standing near each other (Hafner *et al.*, 1982).

The present study showed that, Little Egret prey size varied from 5 to 60 cm of fishes and similar observations were reported by Mukherjee (1972). This species fed in flocks and its plumage is white. White plumaged herons are predominantly focus feeders and this mode of feeding is profitable for Little Egret (Kushlan, 1978). The foraging of Little Egret was observed more in 10 cm to 50 cm water depth. Large Egret's preferred prey length varied between 5 cm to 60 cm.

The estimated prey length of Whiskered Tern varied from 1.2 cm to 5 cm fishes and the species foraged in shallow water and open water habitats.

The flock size attained by Little Egret depends on prey availability (Fasola, 1982; Fasola and Ghidini, 1983; Hafner and Britton, 1983). It also forages with other species of birds like the Little Cormorant, Median Egret, Cattle Egret and other species of wetland birds. This observation in the Kole wetlands are comparable with other reported studies on species like Little Cormorant (Pandey, 1958), Reed Cormorant (Fraser, 1974), Ibis (Morris, 1978), African Spoonbill (Conner, 1979) and Median Egret (Sodhi, 1986). Blacker (1969), Willard (1977), Kushlan (1978), Sodhi and Khera (1984) suggested that the egrets yellow feet might attract prey during food stirring. The egrets and herons at Kole wetlands were found feeding in flocks of size, which varied from 1 to >1000 and were feeding in association with Gulls, Terns, Storks, Ibis and other waders, which is comparable with the report of Willard (1977). Similar results have been reported by many authors (Parks and Bressler, 1963; Reynolds, 1965; Muller *et al.*, 1972).

Ali and Ripley (1983) recorded the feeding of egrets and herons on worms, crustaceans, insects, molluscs, fishes and prawns. Willard (1977) has reported that egrets and herons mostly feed on juvenile eels and large prawns. The result lends support to the possibility that resource partitioning among the mixed species foraging flocks may occur through dietary difference. In the study area, all the species of wading birds were observed foraging in the same location and time. Feeding aggregations were formed early in the morning, just before sunrise. The relative abundance of foraging wetland species and their flocking patterns will likely influence the predator pressure they can exert on selected groups. This also indicates the ability of each species to exploit various prey items across their range (Kushlan, 1978 and Fasola, 1994).

Earlier studies on prey size selection by wetland birds showed that there was pronounced tendency for larger birds to feed on larger prey (Zwarts, 1985), which is comparable with the present result also. Kalejta (1993) also reported on the prey size differentiation by plovers. Goss-Custard (1979) had mentioned that the birds were able to assess the profitability associated with alternative food items. The preference for particular prey species by birds may not always be related to the size of the prey but related to other chemical composition of the prey

(Pulliam, 1980). The preference of wetland birds for a particular prey species is explained by optimal foraging theory (MacArthur and Pianka, 1966; Zwarts, 1985), which states that, birds selected the most profitable size of prey, usually limited by their capacity to handle them. According to Ali and Ripley (1983) locust, grasshopper, crickets, other insects, lizards, field rats, mice, young and sickly birds were the recorded food items of Black-shouldered Kite. Black-shouldered Kite feeding on a migratory species like Wood Sandpiper is not reported so far and it is an addition to the prey items of Black-shouldered Kite.

Four groups of benthic fauna were recorded from the Kole wetlands in different microhabitats *viz.* Insecta, Polycheata, Crustacea and Gastropoda. Of these, Insecta and Polycheata were showed high abundance in all the microhabitats. The availability of food or close to the surface almost certainly plays a key role in determining distribution patterns of these birds. The present study indicated a strong positive correlation between the wader abundance and the benthic fauna availability at the Kole wetlands. The benthic fauna were an important link in the food web of the wetlands, which ultimately decided the wader density. Depending on the season of the year, the species composition of birds varied. Availability of various food sources in different microhabitats was the determining factor, which controlled the seasonal changes of the bird species.

Migratory species showed an increase in population size during November, December and January. Waders feed on a wide variety of foods such as Insecta, Crustacea and Polychaeta. Invertebrates formed the main food of waders. Waders were opportunistic in feeding and fed on suitable items as they are encountered. The size of prey eaten by shorebirds generally increases with the size of the shorebirds. At the Kole wetlands, the abundance of Curlew Sandpiper is closely related to Insecta and Crustacea. Curlew Sandpiper is tactile forager and prey on nerids of different sizes in proportion to their relative abundance in the substratum, avoiding only the smallest nerids (Kalejta, 1993). The absence of prey size selectivity in Curlew Sandpiper implies that their foraging density will be determined by prey abundance, but they should forage preferentially in areas where nerids are large and abundant. The Curlew Sandpiper fed on a wider variety of food items and the importance of benthic fauna in their diet has been reported previously (Kalejta, 1993). Little Stint fed on a wide variety of inter-tidal invertebrates generally Insecta and Crustacea. Feeding Little Stints favoured areas of mud flats, because of the high abundance of benthic fauna.

Population of Little Ringed Plover and Lesser Sand Plover were significantly correlated with Insecta, Polychaeta and Crustacea, which is comparable with the study of Kalejta (1993). Occurrence of wader species like Common Sandpiper was significantly correlated with the abundance of the Crustacea and Polychaeta. Abundance of Curlew was found significantly correlated with Insecta, and negatively correlated with Crustacea and Gastropoda. Wood Sandpiper was found significantly correlated with Insecta Crustacea and Gastropoda.

Population of Common Red Shank was found to be significantly positively correlated with population of Insecta and Crustacea and negatively correlated with Polycheata. Significant correlation was also found between the abundance of Common Green Shank with Gastropoda and Insecta. The population abundance of Green Sandpiper was found significantly negatively correlated with Gastropoda. However, a positive significant correlation was found with Insecta. The abundance of Small Indian Practincole was found to be significantly positively correlated with Polychaeta, Gastropoda, Crustacea and Insecta.

Mud flats greatly influence the distribution, abundance and behaviour of soil organisms (Connors *et al.*, 1981). For waders, mud flats were good site of foraging and availability of prey (Puttick, 1981; Evans and Dugan, 1984). Similar habitat preference was shown in waders at the Berg River Estuary, South Africa (Kalejta and Hockey, 1994); at Iceland (Gudmundsson and Gardarsson, 1993); at Chilika Lake (Acharya, 2000). The high density of benthic fauna in shallow water serves as a foraging habitat of waders, which has also been recorded by Sampath *et al.* (1995) in Great Vedaranyam Salt Swamp and Pichavaram mangroves.

The wader species fed mostly in mud flats, shallow waters and water edge, presumably because many species were more active in the microhabitats. Optimal foraging theory predicts that predators should aggregate in areas where their net rate of energy gain is maximal. In the case of waders, it appears that the numerical abundance of prey is a more important determinant of foraging dispersion than the demography of the prey population. Sediment characteristics and penetrability may also influence the availability of estuarine polychaetes to birds (Myers *et al.*, 1980).

❑❑❑

Chapter 12

Summary and Conclusion

Tropical forest bird community

Species richness at Silent Valley and Mukkali were similar. Most of the birds at Silent Valley were sighted with in a distance of 4m during the census. At Mukkali this distance was up to 9m. Yellowbrowed Bulbul was the most common and dominant species found at Silent Valley. Most common species found at Mukkali was Black Drongo and dominant species was Jungle Babbler. A total of hundred and thirty-seven species of birds were recorded from the study area. There is no significant difference in bird species richness between years during monsoon and summer seasons at Silent Valley. But a significant difference in species richness is obtained between years in both seasons, at Mukkali.

Four endemic and threatened species of birds were recorded from the study area. This shows the conservation value of the study area. Total number of birds sighted in each month showed significant difference at Silent Valley and Mukkali. This is due to the movement of birds influenced by the climatic and other parameters. Overall density of birds was high at Silent Valley (1122 birds/km^2) and low at Mukkali (780 birds/km^2). This indicates the ability of tropical evergreen forest to harbour more birds than tropical moist deciduous forest. The number of individuals/km^2 recorded is comparable to other tropical forests in the world and from this observation we can also conclude that the study area is supporting a rich bird community.

Grey Jungle Fowl, Malabar Whistling Thrush, Southern Tree Pie and Yellowbrowed Bulbul showed stable population, while Black Bulbul, doves and pigeons showed an increase in population during summer season at Silent Valley. The difference in abundance of these two groups, caused due to the local migration may be to cope with the resource availability and climatic conditions. Black Bulbul is a known local migrant. The Grey Junglefowl, the Malabar Whistling Thrush, the Southern Tree Pie and the Yellowbrowed Bulbul showed stable population, while the Black Bulbul, doves and pigeons showed an increase in population during summer season at Silent Valley. The rest could have moved out due to rainfall and shifts in the abundance of prey. The differences in abundance of these two groups, caused due to the local migration may be to cope with the resource availability and climatic conditions. Of these the Black Bulbul is a known local migrant.

Species richness indices were higher at Mukkali than at Silent Valley. Species abundance models follow truncated lognormal distribution at both sites. This clearly explains the existence of an undisturbed and natural, bird community at both the areas. Absence of single dominant species is a feature of tropical forests, which was also found at Silent Valley and Mukkali. It further indicates that a disturbed moist deciduous forest also can hold a diverse bird community similar to evergreen forests. Diversity indices were high at both the places and Mukkali had higher bird diversity. As the microhabitats were diverse at Mukkali it is quite natural that the area is supporting a diverse bird community comparable to Silent Valley. Similarity indices showed that the two areas are similar in bird community only at 40 per cent level. This may be due to the difference in habitat conditions at the two sights. Evenness measures indicate the existence of more rare species at Silent Valley.

As the census of birds were carried in the line transects only, the number of bird taxa recorded may be less compared to the birds found in the whole area. One important observation is that Silent Valley is not a major abode for migrating birds. But compared to other areas like the Periyar Tiger Reserve, Thekkady or Parambikulam wildlife sanctuary, species composition of birds is less at Silent Valley. This may be due to the fact that Silent Valley area being a tropical evergreen forest island, habitat diversity is poor for birds. Number of birds, monthly density and species richness of birds decrease during monsoon season and increase during summer in the study area. Rainfall has significant negative correction on the number of species than on the total number of birds or density at Silent Valley. This is probably because some species avoid the area at the time of extensive rainfall.

Significant difference is obtained between summer and monsoon season in total number of birds and density at Silent Valley. But no such difference is obtained at Mukkali. Compared to Silent Valley rainfall was less at Mukkali. Rainfall has significant negative correlation on the number of birds, density of birds and number of species at both the study areas. But the influence of rainfall on bird community at Mukkali was of low magnitude. There is a drop in the quantum of food in the form of insects, fruit or honey during the months of monsoon. A significant positive correction is obtained between the abundance of flowers and total number of birds. But no significant correlation was obtained between the other types of food and bird abundance. This indicates the continuous availability of food during all the seasons.

Frugivorous birds were more numerous during summer at both Silent Valley and Mukkali. This can be attributed to the fruiting season of trees present in the area. Vertical distribution of bird Community. Species richness of birds was greater at lower and middle levels at both the places. Significant positive correlation obtained between the foliage abundance and total number of birds and number of bird species. This indicated to the positive relationship between bird diversity and foliage abundance in a tropical evergreen and moist deciduous forest. From both the places similar observations were obtained. This also fits well with the theory put forward Mac Arthur's theory for the tropical forests of other countries that bird diversity is linked to the foliage abundance.

In all the height strata omnivorous and insectivorous birds were of equal abundance. Carnivorous birds were more numerous in higher strata at the Silent Valley and Mukkali. Microhabitats like branches of trees and foliage were highly utilized by the arborial birds among the available habitats in the study area. The general overlap in microhabitat use was low in all the habitats. This observation shows the high dependence of tropical forest birds on trees. It also indicates to the diversity in habitat utilization among the tropical forest birds. Foraging ecology studies show gleaning as the principal method used for feeding and this observation also aligns neatly with the observations from other tropical forests. Birds collected their food from seven major substrata. There was never any complete overlap found in any of the foraging parameters studied. This indicated the behavioural adaptations by which, tropical birds avoided competition within the space of a small area.

Generally populations of insectivorous birds dominated both of the study areas followed by omnivorous and frugivorous birds. Frugivorous birds were more abundant at the Silent Valley than at Mukkali. This was to be expected as the Silent Valley had more endemic fruit trees and fruit bearing shrubs. At present the study area have only the status of the reserve forest, no serious efforts are being made to conserve the forests and wildlife. Felling of trees, incidence of fire, collection of minor forest produce (M.F.P.), wildlife poaching and fire wood collection were some of the major types of damages recorded from this area. Eventhough clear felling has been stopped by the Forest Department the felling of eucalyptus plantations adjacent to the study site were conducted in 1990. This has caused severe disturbance in the area for about six months. Workers raised temporary sheds and were staying inside the forest during this period. Many wildlife species, which have been habituating, found in the area have moved to the adjacent places due to the disturbance. Conversion of Eucalyptus plantations into natural forests will be desirable for the bird community.

Another important anthropomorphic influence found in the area was the incidence of intermittent fire. Every year during summer forests got burned. Local people often caused the fire, either purposefully or unknowingly. In 1990 extensive areas have been burnt in this way. Much attention was not paid to control the fire. Main precaution taken against the incidence of fire was the maintenance of fire lines. Apart from this, construction of watchtowers and appointment of firewatchers for monitoring and detection of fire will improve the situation. Collection of minor forest produces like bark of *Cinnamomum* sp., honey, wild cardamom, wild pepper, medicinal plants and cane was another disturbance in the area. Mainly tribal people were engaged in these types of activities. The real menace is not in collecting such materials but due to litting fire for facilitating their collection. In addition to this extraction of cane was another problem recorded from this area. Strict vigilance by the forest staff is the only deterrent for this problem.

Poaching of wild animals for pot was another hazard. Eventhough extensive poaching was not prevalent, small scale poaching was going on in the study site. Poaching of Lion-tailed Monkeys and Nilgiri Langur is in vogue in the area towards Attappady side. The method is by isolating a group of monkeys on a group of trees by removing the nearby trees and leaving only one way to escape. On this path way traps will be placed and monkeys were chased to the net traps. They are used in ayuervedic medicinal preparations or as food. An estate adjacent to this area is another threat to this forest as it takes away the prime forest habitat, which would have been available to the animals.

The study distinctively analyses two tropical bird communities and show its linkage to the habitat, vegetation and climatic parameters. The study compares the bird communities at two sites with different supporting vegetation and shows the variations and similarities in the bird community. The evaluation of the area shows the conservation value of this reserve forest, which is comparable to the world standards and recommends it preservation as a protected area for the posterity.

Tropical wetland bird community

Avifauna of the Kole wetlands was studied from November 1998 to October 2001 to understand the structure, species composition and distribution of birds in the area. One hundred and eighty two taxa of birds were recorded and these belong to 50 Families under 16 Orders. Among the 182 species, 100 were resident, 81 were migratory and one species was a straggler; 48 bird species were newly recorded during this study. Among the migrants, 49 were trans-continental migratory species and 32 species were local migrants.

Habitat wise classification revealed that 25 per cent of species were dependent on aquatic habitats followed by waders (21 per cent) and terrestrial birds (54 per cent). Maximum number of species was insectivores followed by omnivores and aquatic feeders. Nine species of waterfowls were recorded, out of these Garganey, Northern Pintail, Lesser Whistling Teal, Common Teal were abundant, and rest of the species was sighted less than 500 individuals. The sighting of the Black Stork and Lesser Frigate Bird from the Kole wetlands was the first record from the coastal plains and central Kerala. Northern Shoveller was first recorded during the study and this is the second specific report of the species from Kerala.

One vulnerable and five near threatened species were recorded from the Kole wetlands, namely Spot-billed Pelican, Darter, Painted Stork, Oriental White Ibis, Ferruginous Pochard, and Pallid Harrier. Indian Rufous Babbler was an endemic bird species of the Western Ghats reported from the Kole wetlands. Kole wetlands are an ideal habitat for migratory and resident birds, especially for the winter visitors.

During the present study, birds were observed 5,382 times and 4,32,663 birds were counted. Species richness and abundance of birds showed high values in the Kole wetlands, which is comparable to other wetlands in Kerala and in India. Occurrence of 97 species of birds in a month is commendable, which showed the importance of the area for the migratory species. Species richness of avifauna varied in different months and the highest was recorded in December 1999, and the lowest in June 1999. Species richness increased during the migratory season and decreased during the southwest monsoon.

Total number of birds varied from 35 to 8033 individuals in a month and the highest was recorded during November 2000 and the lowest during June 1999. Shannon Index for the whole wetland bird community was 3.11, which indicated the high diversity of birds. Highest density was recorded in the month December (29,158 birds/ha), which was followed by November (24,373 birds/ha). Among the four intensive study areas, species richness and diversity were highest at Parappur. Abundance and density of birds were highest at Kanjany. The highest Diversity Index (H') was observed in the month of June at Parappur (2.91) and the lowest in the month of February at Parappur (1.53).

Highest density was recorded in December (53,994 birds/ha) at Kanjany and lowest in July (103 birds/ha) at Chettupuzha. Among the 82 wetland bird species, Whiskered Tern was highest in dominance followed by Little Egret and Little Cormorant. Species-abundance distribution at the Kole wetlands followed the truncated lognormal model, which indicated the presence of a natural bird community in the area. High similarity was observed between Kanjany and Enamavu, because these two intensive study sites are continuous stretch of the Kole wetlands. Presence of 39 species of waders showed the importance of the wetlands for the migratory birds. The Kole wetlands supported waders similar to the known habitats such as Chilika Lake, Pulicate Lake, Gulf of Mannar and Great Vedaranyam Swamp.

Species richness and abundance of birds were highest in the dry season of 2000 (120) and lowest in the wet-I season of 1999 (46). Diversity index (H′) was highest in the dry season of 2000 (3.19), followed by the dry season of 2001 (2.89). Species richness was highest in the third year migratory season (2000-2001) (127) followed by second year (1999-2000) (119). Highest abundance was in the third year migratory season (2000-2001) (1,88,006) and lowest in the first year migratory season (1998-1999) (19,350).

Species richness and abundance of waders were maximum during the migratory season. Increase in the bird population during the second and third year was due to the early exposure of the preferred microhabitat of waders. The highest number of birds was recorded during November of all the three years, which showed the influx of birds into the region due to trans-continental migration. Abundance of Black-tailed Godwit, Great Knot and Marsh Sandpiper was high in the Kole wetlands compared to the other wetlands of Kerala and other States.

Significant negative correlation was found between the rainfall and the bird community parameters. Highest number of birds was recorded during the replanting season and sowing period of paddy. Significant negative correlation was also obtained between the paddy height and the abundance of birds. During the migratory season, the Kole wetlands supported more waders than the other sites of Kerala in terms of species richness and abundance of birds. The area served many avian species for a wide variety of purposes such as nesting, roosting and wintering ground. The present study showed that the Kole wetlands are one of the important regions in Kerala for winter visitors.

Preferred prey length of the Little Cormorant was 12 cm fishes and the highest number of foraging attempts was recorded in 101 to 150 cm water depth. Indian Pond-Heron preferred fishes of 7 cm and water depth of feeding was less than 10 cm. The Little Egret preferred size of fish was 20 cm and water depth was less than 10 cm. Preferred prey length of Large Egret was fishes of 30 cm length and water depth for foraging ranged between 11 to 20 cm. The preferred prey length of the Whiskered Tern was fishes of 2.5 cm length and foraging water depth was between 101 to 150 cm.

Mud flats greatly influence the distribution and abundance of soil organisms and mud flats were good site for foraging due to the availability of prey. The study indicated a strong positive correlation between abundance of waders and benthic fauna in the Kole wetlands. Waders fed on a wide variety of groups such as Insecta, Crustacea, Gastropoda and Polychaeta and the abundance of Curlew Sandpiper is closely related to Insecta and Crustacea.

The population of Little Ringed Plover and Lesser Sand Plover was significantly correlated with Insecta, Polychaeta and Crustacea. Species like Common Sandpiper were significantly correlated with the abundance of the Crustacea and Polychaeta. The abundance of Curlew was found significantly positively correlated with Insecta, and negatively correlated with Crustacea and Gastropoda. The Wood Sandpiper was found significantly correlated with Insecta, Crustacea and Gastropoda.

The population of Common Red Shank was found to be significantly positively correlated with population of Insecta and Crustacea, negatively correlated with Polycheata. The abundance of Common Green Shank was found significantly correlated with Gastropoda and Insecta. The population abundance of Green Sandpiper was significantly negatively correlated with Gastropoda. The abundance of Small Practincole was found significantly correlated with Polychaeta, Gastropoda, Crustacea and Insecta. Food and feeding behavior studies showed that sufficient prey was available in the Kole wetlands and the selection of food in the wetland birds depended on the difference in the feeding technique, body size and bill morphology.

Twelve microhabitats were identified from the Kole wetlands, namely the floating vegetation, bund, bamboo pole, trees, electric line, water edge, shallow water, mud flats, grass, paddy fields, open water and harvested paddy fields. Among the twelve microhabitats, shallow water was highly preferred. Species richness was highest on the electric line (74) followed by mud flats (69). The highest diversity index was recorded on trees (3.44) and the lowest on bunds (1.38).

The highest species density was recorded in the shallow water (67,788 birds/ha) followed by the paddy fields (56,774 birds/ha) and mud flats (56,308 birds/ha). The highest niche breadth was recorded for the Red-wattled Lapwing and the Indian Pond-Heron. Mean species richness was significantly high on the electric lines during the dry season, wet-I season and in the mud flats during the wet-II season. Diversity Index (H') was significantly high on trees in the dry and the wet-II season and during the wet-I season diversity index was high on the electric lines. Density of birds was high in the paddy fields during the dry and wet-II seasons and significantly high in the floating vegetation during the wet-II season.

The variation in the species richness, diversity and density of birds in different microhabitats were clear indication of their habitat preference. The appearance of mud flats attracted large number of waders during the migratory season. The distribution of birds depended on the abundance of food and habitat structure. Vast extent of mudflats available at the Kole wetlands was the prime habitat for waders. The habitat selection of wetland birds in the Kole wetlands was influenced by the availability of prey and accessibility, where as the water depth influenced the accessibility of the prey to the birds.

A total of 145 men and 10 women respondents were interviewed for collecting the information on the conservation of bird community. Majority of local people opinioned that, individuals from out side were responsible for poaching. Poaching of birds was primarily for food and for sale by the professional hunters and as a leisure time activity. Poaching of birds was the major identified problem, which is detrimental to the wetland birds. Shooting and poisoning were the main methods employed for killing the birds. Poisoning was mainly targeted towards Indian Pond-Heron, Egrets and Black-crowned Night-Heron. Burning of grass in the dykes was another practice, which is harmful to the bird populations. Nestlings and eggs of resident birds were destroyed due to this practice. Mud flats of the Kole wetlands have been severely degraded by burning, drainage and clearance due to clay mining and agricultural activities.

Important conservation problems identified from the area were habitat alteration, poaching, fire and fishing. Kole wetlands were reclaimed for various purposes like raising coconut plantations and for constructing buildings. As the Kole wetlands are serving as "Stepping stone" for the trans-continental migrants, urgent measures are needed to protect this wetland ecosystem for the conservation of migratory birds. As this wetland is coming under the 'Central Asian - Indian flyway' protection of migratory bird species is of the highest priority.

❑❑❑

Appendix

Appendix 1. List of birds recorded from the Silent Valley National Park

Sl. No.	Common Name	Scientific Name	Status
	Charadriformes		
	Ardeidae		
1.	Indian Pond-Heron	*Ardeola grayii* (Sykes)	R
	Falconiformes		
	Accipitridae		
2.	Black-shouldered Kite	*Elanus caeruleus* (Desfontaines)	R
3.	Shikra	*Accipiter badius* (Gmelin)	R
4.	Black Eagle	*Ictinaetus malayensis* (Temminck)	R
5.	Greater Grey-headed Fish-Eagle	*Icthyophaga ichthyaetus* (Horsfield)	R
6.	Short-toed Snake-Eagle	*Circaetus gallicus* (Gmelin)	R
7.	Crested Serpent-Eagle	*Spilornis cheela* (Latham)	R
8.	Brahminy Kite	*Haliastur indus* (Boddert)	R
	Falconidac		
9.	Peregrine Falcon	*Falco peregrinus* (Tunstall)	R

Contd...

Appendix 1. Contd...

	Galliformes		
	Phasianidae		
10.	Painted Bush-Quail	*Perdicula erythrorhyncha* (Sykes)	R
11.	Red Spurfowl	*Galloperdix spadicea* (Gmelin)	R
12.	Grey Junglefowl	*Gallus sonneratii* Temminck	R
	Turnicidae		
13.	Common Buttonquail	*Turnix suscitator* (Gmelin)	R
	Columbiformes		
	Columbidae		
14.	Yellow-legged Green-Pigeon	*Treron phoenicoptera* (Latham)	LM
15.	Pombadour Green-Pigeon	*Treron pompadora* (Gmelin)	R
16.	Mountain Imperial-Pigeon	*Ducula badia* (Raffles)	R
17.	Green Imperial Pigeon	*Ducula aenea* (Linnaeus)	R
18.	Nilgiri Wood-Pigeon	*Columba elphinstonii* (Sykes)	R
19.	Spotted Dove	*Streptopelia chinensis* (Scopoli)	R
20.	Emerald Dove	*Chalcophaps indica* (Linnaeus)	R
21.	Blue Rock Pigeon	*Columba livia* (Gmelin)	R
	Psittaciformes		
	Psittacidae		
22.	Blue-winged Parakeet	*Psittacula columboides* (Vigors)	R
23.	Blossom-headed Parakeet	*Psittacula cyanocephala* (Linnaeus)	R
24.	Indian Hanging-Parrot	*Loriculus vernalis* (Sparrman)	R
25.	Rose-ringed Parakeet	*Psittacula krameri* (Scopoli)	R
	Cuculiformes		
	Cuculidae		
26.	Red-winged Crested Cuckoo	*Clamator coromandus* (Linnaeus)	M
27.	Brainfever Bird	*Hierococcyx varius* (Vahl)	R
28.	Asian Koel	*Eudynamus scolopacea* (Linnaeus)	R
29.	Greater Pheasant	*Centropus sinensis* (Stephens)	R
	Strigiformes		
	Strigidae		
30.	Collared Scops-Owl	*Otus bakkamoena* Pennant	R
31.	Forest Eagle-owl	*Bubo nipalensis* Hodgson	M
32.	Jungle Owlet	*Glaucidium radiatum* (Tickell)	R
33.	Short-eared Owl	*Asio flammeus* (Pontoppidan)	R
34.	Brown Fish-Owl	*Ketupa zeylonensis* (Temminck)	R
35.	Brown Hawk-Owl	*Ninox scutulata* (Raffles)	R
36.	Spotted Owlet	*Athene brama* (Temminck)	R
37.	Brown Wood-Owl	*Strix leptogrammica* Temminck	R
	Camprimulgiformes		
	Caprimulgidae		
38.	Indian Jungle Nightjar	*Caprimulgus indicus* Latham	R
	Apodidae		
39.	Brown-backed Needletail-Swift	*Hirundapus giganteus* (Tamminck)	R

Contd...

Appendix 1. Contd...

40.	White-rumped Needletail-Swift	*Zoonavena sylvatica* (Tickell)	R
	Trogoniformes		
	Trogonidae		
41.	Malabar Trogon	*Harpactes fasciatus* (Pennant)	R
	Alcedinidae		
42.	White-breasted Kingfisher	*Halcyon smyrnensis* (Smyrnensis)	R
	Meropidae		
43.	Chestnut-headed Bee-eater	*Merops leschenulti* Vieillot	R
	Upupidae		
44.	Common Hoopoe	*Upupa epops* (Linnaeus)	R
	Bucerotidae		
45.	Malabar Grey Hornbill	*Ocyceros griseus* (Latham)	R
46.	Great Pied Hornbill	*Buceros bicornis* Hodgson	R
	Capitonidae		
47.	White-cheeked Barbet	*Megalaima viridis* (Boddaert)	R
	Picidae		
48.	Small Yellow-naped Woodpecker	*Picus chlorolophus* Vieillot	R
49.	Common Golden-backed Woodpecker	*Dinopium javanense* Ljungh	R
50.	Lesser Golden-backed Woodpecker	*Dinopium benghalense* Linnaeus	R
51.	Great Black Woodpecker	*Dryocopus javensis* (Horsfield)	R
52.	Heart-spotted Woodpecker	*Hemicircus canente* (Lesson)	R
53.	Greater Golden-backed Woodpecker	*Chrysocolaptes lucidus* Scopoli	R
54.	Little Scaly-bellied Green Woodpecker	*Picus xanthopygaeus* (J.E. Gray & G.R. Gray)	R
	Pittidae		
55.	Indian Pitta	*Pitta brachyura* (Linnaeus)	M
	Hirundinidae		
56.	Dusky Crag-Martin	*Hirundo concolor* Sykes	R
57.	House Swallow	*Hirundo tahitica* Gmelin	R
58.	Red-rumped Swallow	*Hirundo daurica* Linnaeus	R
	Lanidae		
59.	Rufous-backed Shrike	*Lanius schach* Linnaeus	R
60.	Bay-backed Shrike	*Lanius vittatus* Valenciennes	M
61.	Greater Grey Shrike	*Lanius excubitor*	R
	Oriolidae		
62.	Eurasian Golden Oriole	*Oriolus oriolus* Linnaeus	R
63.	Black-headed Oriole	*Oriolus xanthornus* (Linnaeus)	R
64.	Black-naped Oriole	*Oriolus chinensis* Linnaeus	M
	Dicruridae		
65.	Black Drongo	*Dicrurus macrocercus* Vieillot	R
66.	Bronzed Drongo	*Dicrurus aeneus* Vieillot	R

Contd...

Appendix 1. Contd...

67.	Greater Racket-tailed Drongo	*Dicrurus paradiseus* (Linnaeus)	R
68.	Ashy Drongo	*Dicrurus leucophaeus* Vieillot	R
69.	Spangled Drongo	*Dicrurus hottentottus* (Linnaeus)	R
70.	White-bellied Drongo	*Dicrurus caerulescens* (Linnaeus)	R
	Sturnidae		
71.	Common Myna	*Acridotheres tristis* (Linnaeus)	R
72.	Grey-headed Myna	*Sturnus malabaricus* (Gmelin)	LM
73.	Jungle Myna	*Acridotheres fuscus* (Wagler)	R
74.	Common Hill-Myna	*Gracula religiosa* Linnaeus	R
	Corvidae		
75.	White-bellied Tree Pie	*Dendrocitta leucogastra* Gould	R
77.	Indian Treepie	*Dendrocitta vagabunda* (Latham)	R
76.	House Crow	*Corvus splendens* Vieillot	R
78.	Jungle Crow	*Corvus macrcorhynchos* Wagler	R
	Campehiagidae		
79.	Scarlet Minivet	*Pericrocotus flammeus* (Forster)	R
80.	Cuckoo Shrike	*Coracina sp.*	R
81.	Black-winged Cuckoo-Shrike	*Coracina melaschistos* (Hodgson)	M
82.	Pied Flycatcher-Shrike	*Hemipus picatus* (Sykes)	R
83.	Large Woodshrike	*Tephrodornis gularis* (Raffles)	R
	Irenidae		
84.	Asian Fairy-Bluebird	*Irena puella* (Latham)	R
85.	Common Iora	*Aegithina tiphia* (Marshall)	R
86.	Gold-fronted Chloropsis	*Chloropsis aurifrons* (Temminck)	R
87.	Jerdon's Chloropsis	*Chloropsis cochinchinensis* (Gmelin)	R
	Pycnonotidae		
88.	Black-crested Bulbul	*Pycnonotus melanicterus* (Gmelin)	R
89.	Red-whiskered Bulbul	*Pycnonotus jocosus* (Linnaeus)	R
90.	Red-vented Bulbul	*Pycnonotus cafer* (Linnaeus)	R
91.	Yellow-browed Bulbul	*Iole indica* (Jerdon)	R
92.	Black Bulbul	*Hypsipetes leucocephalus* (P.L.S. Muller)	R
	Muscicapidae		
93.	Quaker Tit-Babbler	*Alcippe poioicephala* (Jerdon)	R
94.	Spotted Babbler	*Pellorneum ruficeps* Jerdon	R
95.	Hodgson's Scimitar-Babbler	*Pomatorhinus schisticeps* Harington	R
96.	Black-headed Babbler	*Rhopocichla atriceps* (Jerdon)	R
97.	Rufous-bellied Babbler	*Dumetia hyperythra* (Franklin)	R
98.	Indian Rufous Babbler	*Turdoides subrufus* (Sharpe)	R
99.	Jungle Babbler	*Turdoides striatus* (Dumont)	R
100.	White-headed Babbler	*Turdoides affinis* (Jerdon)	R
101.	Laughing Thrush	*Garrulax delesserti* (Jerdon)	R
102.	Blue-throated Flycatcher	*Cyornis rubeculoides* (Vigors)	R
103.	Tickell's Blue-Flycatcher	*Cyornis tickelliae* Blyth	R

Contd...

Appendix 1. Contd...

104.	Brown-breasted Flycatcher	*Muscicapa muttui* (Layard)	M
105.	Verditer Flycatcher	*Eumyias thalassiana* (Swainson)	M
106.	Rusty-tailed Flycatcher	*Muscicapa ruficaudata* Swainson	M
107.	Asian Paradise-Flycatcher	*Terpsiphone paradisi* (Linnaeus)	R
108.	Asian Brown Flycatcher	*Muscicapa dauurica* Pallas	R
109.	White-bellied Blue-Flycatcher	*Cyonis pallipes* (Jerdon)	R
110.	Black-and-Orange Flycatcher	*Fiedula nigrorufa* (Jerdon)	R
111.	Grey-headed Flycatcher	*Culicicapa ceylonensis* (Swainson)	R
112.	Black-naped Monarch-Flycatcher	*Hypothymis azurea* (Boddaert)	R
113.	Streaked Fantail-Warbler	*Cisticola junciidis* Rafinesque	R
114.	Francklin's Prinia	*Pirinia hodgsonii* Blyth	R
115.	Common Tailorbird	*Orthotomus sutorius* (Pennant)	R
117.	Large-billed Leaf-Warbler	*Phylloscopus magnirostris* Blyth	M
118.	Greenish Leaf-Warbler	*Phylloscopus trochiloides* (Sundevall)	M
119.	Western Crowned Warbler	*Phylloscopus occipitalis* Blyth	M
116.	Oriental Magpie-robin	*Copsychus saularis* Linnaeus	R
120.	White-throated Ground Thrush	*Zoothera citrina cyanotus* (Jardine & Selby)	R
121.	Orange-headed Thrush	*Zoothera citrina* (Latham)	R
122.	Pied Bushchat	*Saxicola caprata* (Linnaeus)	R
123.	Malabar Whistling-Thrush	*Myiophonus horsfieldii* (Vigors)	R
124.	Blue Rock-Thrush	*Monticola solitarius* (Linnaeus)	M
125.	Eurasian Blackbird	*Turdus merula* Linnaeus	R
	Paridae		
126.	Great Tit	*Parus major* Linnaeus	R
127.	Black-lored Tit	*Parus xanthogenys* Vigors	R
	Motacillidae		
128.	Velvet-fronted Nuthatch	*Sitta frontalis* Swainson	R
129.	Oriental Tree Pipit	*Anthus hodgsoni* Richmond	M
130.	Paddyfield Pipit	*Anthus rufulus* Vieiilot	M
131.	Forest Wagtail	*Dendronanthus indicus* (Gmelin)	M
132.	Yellow Wagtail	*Motacilla flava* Linnaeus	M
133.	Grey Wagtail	*Motacilla cinera* Tunstall	M
134.	Large Pied Wagtail	*Motacilla maderaspatensis* Gmelin	R
135.	White Wagtail	*Motacilla alba* Linnaeus	M
	Dicaeidae		
136.	Thick-billed Flowerpecker	*Dicaeum agile* (Tickell)	R
137.	Tickell's Flowerpecker	*Dicaeum erythrorhynchos* (Latham)	R
	Nectariniidae		
138.	Purple-rumped Sunbird	*Nectarina zeylonica* (Linnaeus)	R
139.	Small Sunbird	*Nectarina minima* (Sykes)	R
140.	Purple Sunbird	*Nectarina asiatica* (Latham)	R
141.	Loten's Sunbird	*Nectarina lotenia* Linnaeus	R

Contd...

Appendix 1. Contd...

142.	Little Spiderhunter	*Arachnothera longirostris* (Latham)	R
	Zosteropidae		
143.	Oriental White-eye	*Zosterops palpebrosus* (Temminck)	R
	Ploceidae		
144.	House Sparrow	*Passer domesticus* (Linnaeus)	R
145.	Yellow-throated Sparrow	*Petronis xanthocollis* (Burton)	R
146.	White-throated Munia	*Lonchura malabarica* (Linnaeus)	R
147.	Black-throated Munia	*Lonchura kelaarti* (Jerdon)	R
148.	Black-headed Munia	*Lonchura mallacca* (Linnaeus)	R

R = Resident, M = Migrant, LM = Local Migrant

Appendix 2. List of birds recorded from the Kole wetlands

Sl. No.	Common name	Scientific name	Status
	Podicipediformes		
	Podicipedidae		
1.	Little Grebe	*Tachybaptus ruficollis* (Pallas)	R
	Pelecaniformes		
	Pelecanidae		
2.	Spot-billed Pelican	*Pelecanus philippensis* Gmelin	R
	Sulidae		
3.	Masked Booby	*Sula dactylatra* Lesson	M
	Phalacrocoracidae		
4.	Little Cormorant	*Phalacrocorax niger* (Vieillot)	R
5.	Great Cormorant	*Phalacrocorax carbo* (Linnaeus)	LM
6.	Indian Shag	*Phalacrocorax fuscicollis* Stephens	LM
	Anhingidae		
7.	Darter	*Anhinga melanogaster* Pennant	LM
	Fregatidae		
8.	Lesser Frigatebird	*Fregata ariel* (G.R. Gray)	S
	Ciconiiformes		
	Ardeidae		
9.	Grey Heron	*Ardea cinerea* Linnaeus	R
10.	Purple Heron	*Ardea purpurea* Linnaeus	R
11.	Little Green Heron	*Butorides striatus* (Linnaeus)	R
12.	Indian Pond-Heron	*Ardeola grayii* (Sykes)	R
13.	Cattle Egret	*Bubulcus ibis* (Linnaeus)	R
14.	Large Egret	*Casmerodius albus* (Linnaeus)	R
15.	Median Egret	*Mesophoyx intermedia* (Wagler)	R
16.	Little Egret	*Egretta garzetta* (Linnaeus)	R
17.	Western Reef-Egret	*Egretta gularis* (Bosc)	R
18.	Black-crowned Night- Heron	*Nycticorax nycticorax* (Linnaeus)	R
19.	Chestnut Bittern	*Ixobrychus cinnamomeus* (Gmelin)	R
20.	Yellow Bittern	*Ixobrychus sinensis* (Gmelin)	R
21.	Black Bittern	*Dupetor flavicollis* (Latham)	R
	Ciconiidae		
22.	Painted Stork	*Mycteria leucocephala* (Pennant)	LM
23.	Asian Openbill-Stork	*Anastomus oscitans* (Boddaert)	LM
24.	White-necked Stork	*Ciconia episcopus* (Boddaert)	R
25.	Europian White Stork	*Ciconia ciconia* (Linnaeus)	M
26.	Black Stork	*Ciconia nigra* (Linnaeus)	M
	Threskiornithidae		
27.	Oriental White Ibis	*Threskiornis melanocephalus* (Latham)	LM
28.	Black Ibis	*Pseudibis papillosa* (Temminck)	R

Contd...

Appendix 2. Contd...

29.	Eurasian Spoonbill	*Platalea leucorodia* Linnaeus	LM
	Anseriformes		
	Anatidae		
30.	Lesser Whistling-Duck	*Dendrocygna javanica* (Horsfield)	R
31.	Common Teal	*Anas crecca* Linnaeus	M
32.	Northern Pintail	*Anas acuta* Linnaeus	M
33.	Spot-billed Duck	*Anas poecilorhyncha* J.R. Forester	LM
34.	Gadwall	*Anas strepera* Linnaeus	M
35.	Garganey	*Anas querquedula* Linnaeus	M
36.	Northern Shoveller	*Anas clypeata* Linnaeus	M
37.	Ferruginous Pochard	*Aythya nyroca* (Guldenstadt)	M
38.	Cotton Teal	*Nettapus coromandelianus* (Gmelin)	LM
	Falconiformes		
	Accipitridae		
39.	Black-shouldered Kite	*Elanus caeruleus* (Desfontaines)	R
40.	Black Kite	*Milvus migrans* (Boddaert)	R
41.	Brahminy Kite	*Haliastur indus* (Boddaert)	R
42.	Shikra	*Accipiter badius* (Gmelin)	R
43.	Eurasian Sparrowhawk	*Accipiter nisus* (Linnaeus)	M
44.	Oriental Honey-Buzzard	*Pernis ptilorhynchus* (Temminck)	LM
45.	Pallid Harrier	*Circus macrourus* (S.G. Gmelin)	M
46.	Pied Harrier	*Circus melanoleucos* (Pennant)	M
47.	Western Marsh-Harrier	*Circus aeruginosus* (Linnaeus)	M
	Pandionidae		
48.	Osprey	*Pandion haliaetus* (Linnaeus)	R
	Galliformes		
	Phasianidae		
49.	Grey Francolin	*Francolinus pondicerianus* (Gmelin)	R
50.	Red Spurfowl	*Galloperdix spadicea* (Gmelin)	R
51.	Indian Peafowl	*Pavo cristatus* Linnaeus	R
	Gruiformes		
	Rallidae		
52.	Ruddy-breasted Crake	*Porzana fusca* (Linnaeus)	R
53.	Slaty-legged Crake	*Rallina eurizonoides* (Lafresnaye)	R
54.	White-breasted Waterhen	*Amaurornis phoenicurus* (Pennant)	R
55.	Watercock	*Gallicrex cinerea* (Gmelin)	M
56.	Common Moorhen	*Gallinula chloropus* (Linnaeus)	R
57.	Purple Moorhen	*Porphyrio porphyrio* (Linnaeus)	R
58.	Common Coot	*Fulica atra* Linnaeus	LM
	Charadriiformes		
	Jacanidae		
59.	Pheasant-tailed Jacana	*Hydrophasianus chirurgus* (Scopoli)	LM

Contd...

Appendix 2. Contd...

60.	Bronze-winged Jacana	*Metopidius indicus* (Latham)	R
	Rostratulidae		
61.	Greater Painted-Snipe	*Rostratula benghalensis* (Linneaus)	R
	Charadriidae		
62.	Red-wattled Lapwing	*Vanellus indicus* (Boddaert)	R
63.	Pacific Golden-Plover	*Pluvialis fulva* (Gmelin)	M
64.	Lesser Sand Plover	*Charadrius mongolus* Pallas	M
65.	Little Ringed Plover	*Charadrius dubius* Scopoli	M
66.	Kentish Plover	*Charadrius alexandrinus* Linnaeus	LM
	Scolopacidae		
67.	Common Snipe	*Gallinago gallinago* (Linnaeus)	M
68.	Pintail Snipe	*Gallinago stenura* (Bonaparte)	M
69.	Eurasian Woodcock	*Scolopax rusticola* Linnaeus	LM
70.	Black-tailed Godwit	*Limosa limosa* (Linnaeus)	M
71.	Bar-tailed Godwit	*Limosa lapponica* (Linnaeus)	M
72.	Whimbrel	*Numenius phaeopus* (Linnaeus)	M
73.	Eurasian Curlew	*Numenius arquata* (Linnaeus)	M
74.	Common Redshank	*Tringa totanus* (Linnaeus)	M
75.	Marsh Sandpiper	*Tringa stagnatilis* (Bechstein)	M
76.	Common Greenshank	*Tringa nebularia* (Gunner)	M
77.	Green Sandpiper	*Tringa ochropus* Linnaeus	M
78.	Wood Sandpiper	*Tringa glareola* Linnaeus	M
79.	Terek Sandpiper	*Xenus cinereus* (Guldenstadt)	M
80.	Common Sandpiper	*Actitis hypoleucos* Linnaeus	M
81.	Ruddy Turnstone	*Arenaria interpres* (Linnaeus)	M
82.	Great Knot	*Calidris tenuirostris* (Horsfield)	M
83.	Dunlin	*Calidris alpina* (Linnaeus)	M
84.	Curlew Sandpiper	*Calidris ferruginea* (Pontoppidan)	M
85.	Sanderling	*Calidris alba* (Pallas)	M
86.	Little Stint	*Calidris minuta* (Leisler)	M
87.	Temminck's Stint	*Calidris temminckii* (Leisler)	M
88.	Broad-billed Sandpiper	*Limicola falcinellus* (Pontoppidan)	M
89.	Ruff	*Philomachus pugnax* (Linnaeus)	M
	Recurvirostridae		
90.	Black-winged Stilt	*Himantopus himantopus* (Linnaeus)	LM
91.	Pied Avocet	*Recurvirostra avosetta* Linnaeus	M
	Glareolidae		
92.	Small Pratincole	*Glareola lactea* Temminck	LM
	Laridae		
93.	Yellow-legged Gull	*Larus cachinnans* Pallas	M
94.	Brown-headed Gull	*Larus brunnicephalus* Jerdon	M
95.	Black-headed Gull	*Larus ridibundus* Linnaeus	M

Contd...

Appendix 2. Contd...

96.	Whiskered Tern	*Chlidonias hybridus* (Pallas)	LM
97.	Caspian Tern	*Sterna caspia* Pallas	M
	Columbiformes		
	Columbidae		
98.	Blue Rock Pigeon	*Columba livia* Gmelin	R
99.	Spotted Dove	*Streptopelia chinensis* (Scopoli)	R
100.	Eurasian Collared-Dove	*Streptopelia decaocto* (Frivaldszky)	R
	Psittaciformes		
	Psittacidae		
101.	Rose-ringed Parakeet	*Psittacula krameri* (Scopoli)	R
102.	Plum-headed Parakeet	*Psittacula cyanocephala* (Linnaeus)	LM
	Cuculiformes		
	Cuculidae		
103.	Pied Crested Cuckoo	*Clamator jacobinus* (Boddaert)	LM
104.	Brainfever Bird	*Hierococcyx varius* (Vahl)	LM
105.	Indian Cuckoo	*Cuculus micropterus* Gould	R
106.	Banded Bay Cuckoo	*Cacomantis sonneratii* (Latham)	LM
107.	Asian Koel	*Eudynamys scolopacea* (Linnaeus)	R
108.	Greater Coucal	*Centropus sinensis* (Stephens)	R
	Strigiformes		
	Tytonidae		
109.	Barn Owl	*Tyto alba* (Scopoli)	R
	Strigidae		
110.	Spotted Owlet	*Athene brama* (Temminck)	R
111.	Mottled Wood-Owl	*Strix ocellata* (Lesson)	R
	Apodiformes		
	Apodidae		
112.	Alpine Swift	*Tachymarptis melba* (Linnaeus)	R
113.	House Swift	*Apus affinis* (J.E. Gray)	R
114.	Asian Palm-Swift	*Cypsiurus balasiensis* (J.E. Gray)	R
	Coraciiformes		
	Alcedinidae		
115.	Lesser Pied Kingfisher	*Ceryle rudis* (Linnaeus)	R
116.	Small Blue Kingfisher	*Alcedo atthis* (Linnaeus)	R
117.	Stork-billed Kingfisher	*Halcyon capensis* (Linnaeus)	R
118.	White-breasted Kingfisher	*Halcyon smyrnensis* (Linnaeus)	R
119.	Black-capped Kingfisher	*Halcyon pileata* (Boddaert)	LM
	Meropidae		
120.	Blue-tailed Bee-eater	*Merops philippinus* Linnaeus	LM
121.	Small Bee-eater	*Merops orientalis* Latham	R
	Coraciidae		
122.	Indian Roller	*Coracias benghalensis* (Linnaeus)	R

Contd...

Appendix 2. Contd...

	Upupidae		
123.	Common Hoopoe	*Upupa epops* Linnaeus	R
	Piciformes		
	Capitonidae		
124.	White-cheeked Barbet	*Megalaima viridis* (Boddaert)	R
	Picidae		
125.	Lesser Golden-backed Woodpecker	*Dinopium benghalense* (Linnaeus)	R
	Passeriformes		
	Hirundinidae		
126.	Common Swallow	*Hirundo rustica* Linnaeus	LM
127.	House Swallow	*Hirundo tahitica* Gmelin	R
128.	Red-rumped Swallow	*Hirundo daurica* Linnaeus	LM
	Motacillidae		
129.	Paddyfield Pipit	*Anthus rufulus* Vieillot	LM
130.	Eurasian Tree Pipit	*Anthus trivialis* (Linnaeus)	M
131.	Yellow Wagtail	*Motacilla flava* Linnaeus	LM
132.	Citrine Wagtail	*Motacilla citreola* Pallas	M
133.	Grey Wagtail	*Motacilla cinerea* Tunstall	M
134.	Large Pied Wagtail	*Motacilla maderaspatensis* Gmelin	R
	Pycnonotidae		
135.	Red-whiskered Bulbul	*Pycnonotus jocosus* (Linnaeus)	LM
136.	Red-vented Bulbul	*Pycnonotus cafer* (Linnaeus)	R
	Irenidae		
137.	Common Iora	*Aegithina tiphia* (Linnaeus)	R
138.	Gold-fronted Chloropsis	*Chloropsis aurifrons* (Temminck)	R
139.	Jerdon's Chloropsis	*Chloropsis cochinchinensis* (Gmelin)	R
	Laniidae		
140.	Brown Shrike	*Lanius cristatus* Linnaeus	M
	Turdinae		
141.	Oriental Magpie-Robin	*Copsychus saularis* (Linnaeus)	R
142.	Indian Robin	*Saxicoloides fulicata* (Linnaeus)	R
143.	Pied Bushchat	*Saxicola caprata* (Linnaeus)	R
144.	Desert Wheatear	*Oenanthe deserti* (Temminck)	M
	Timaliinae		
145.	White-headed Babbler	*Turdoides affinis* (Jerdon)	R
146.	Common Babbler	*Turdoides caudatus* (Dumont)	R
147.	Jungle Babbler	*Turdoides striatus* (Dumont)	R
148.	Indian Rufous Babbler	*Turdoides subrufus* (Jerdon)	R
	Sylviinae		
149.	Streaked Fantail-Warbler	*Cisticola juncidis* (Rafinesque)	R
150.	Franklin's Prinia	*Prinia hodgsonii* Blyth	R

Contd...

Appendix 2. Contd...

151.	Plain Prinia	*Prinia inornata* Sykes	R
152.	Ashy Prinia	*Prinia socialis* Sykes	R
153.	Common Tailorbird	*Orthotomus sutorius* (Pennant)	R
154.	Indian Great Reed-Warbler	*Acrocephalus stentoreus* (Hemprich & Ehrenberg)	R
155.	Blyth's Reed-Warbler	*Acrocephalus dumetorum* Blyth	LM
	Monarchinae		
156.	Asian Paradise-Flycatcher	*Terpsiphone paradisi* (Linnaeus)	LM
	Dicaeidae		
157.	Tickell's Flowerpecker	*Dicaeum erythrorhynchos* (Latham)	R
158.	Thick-billed Flowerpecker	*Dicaeum agile* (Tickell)	R
	Nectariniidae		
159.	Purple-rumped Sunbird	*Nectarinia zeylonica* (Linnaeus)	R
160	Purple Sunbird	*Nectarinia asiatica* (Latham)	R
161.	Loten's Sunbird	*Nectarinia lotenia* (Linnaeus)	R
	Estrildidae		
162.	Red Munia	*Amandava amandava* (Linnaeus)	R
163.	White-rumped Munia	*Lonchura striata* (Linnaeus)	R
164.	White-throated Munia	*Lonchura malabarica* (Linnaeus)	R
165.	Black-throated Munia	*Lonchura kelaarti* (Jerdon)	R
166.	Spotted Munia	*Lonchura punctulata* (Linnaeus)	R
167.	Black-headed Munia	*Lonchura malacca* (Linnaeus)	R
	Ploceinae		
168.	Black-breasted Weaver	*Ploceus benghalensis* (Linnaeus)	R
169.	Baya Weaver	*Ploceus philippinus* (Linnaeus)	R
170.	Streaked Weaver	*Ploceus manyar* (Horsfield)	R
	Sturnidae		
171.	Common Myna	*Acridotheres tristis* (Linnaeus)	R
172.	Jungle Myna	*Acridotheres fuscus* (Wagler)	R
173.	Grey-headed Starling	*Sturnus malabaricus* (Gmelin)	R
	Oriolidae		
174.	Eurasian Golden Oriole	*Oriolus oriolus* (Linnaeus)	LM
175.	Black-headed Oriole	*Oriolus xanthornus* (Linnaeus)	R
	Dicruridae		
176.	Black Drongo	*Dicrurus macrocercus* Vieillot	R
177.	Ashy Drongo	*Dicrurus leucophaeus* Vieillot	LM
178.	White-bellied Drongo	*Dicrurus caerulescens* (Linnaeus)	R
	Artamidae		
179.	Ashy Woodswallow	*Artamus fuscus* Vieillot	R
	Corvidae		
180.	Indian Treepie	*Dendrocitta vagabunda* (Latham)	R
181.	House Crow	*Corvus splendens* Vieillot	R
182.	Jungle Crow	*Corvus macrorhynchos* Wagler	R

R = Resident, M = Migrant, LM = Local Migrant

□□□

References

Acharya, S. 2000. Studies on some ecological aspects of shorebirds (Family: Charadriidae) with species reference to Plovers of Nalabana Island, Chilika Lagoon, Orissa. Ph.D. Thesis, Utkal University, Bhubaneswar, Orissa. 177 p.

Acharya, S. and S.K. Kar 1996. Checklist of Waders (Charadriiformes) in Chilika Lake, Orissa. *Newsl. Birdwatchers* 36(5): 89-90.

Alatalo, R.V. 1981. Problems in the measurement of evenness in ecology. *Oikos* 37: 199-204.

Alfred, J.R.B., A. Kumar, P.C. Tak and J.P. Sati 2000. Waterbirds of Northern India. *Rec. Zool. Surv. India. Occ. Paper* 190: 1-227.

Ali, S. 1969. The Birds of Kerala. Oxford University Press, Bombay, 444 p.

Ali, S. and H. Whistler 1935-37. The ornithology of Travancore and Cochin. *J. Bombay nat. Hist. Soc.* 8 Parts. 37-39.

Ali, S. and S. D. Ripley 1983. Hand book of the birds of India and Pakistan. Oxford University Press, Oxford. 737 p.

Ali, S. and S.D. Ripley 1983a. A Pictorial Guide to the Birds of the Indian Subcontinent, Oxford University Press, Bombay, 177 p.

Altman, J. 1974. Observational study of behaviour: sampling methods. *Behaviour* 49: 227-267.

Ambedkar, V.C. 1985. Occurrence of the Sandwich Tern (*Sterna sandvicensis*) in India - a ring recovery. *J. Bombay nat. Hist. Soc.* 82(2): 410.

Anon. 1992. Birds of Kole wetlands: A survey report-I. Nature Education Society Thrissur (NEST), in collaboration with Kerala Forest Research Institute (KFRI) and Kerala Forest Department. 16 p.

Anon. 1993. Birds of Kole wetlands: A Survey Report-II. Nature Education Society Thrissur (NEST), in collaboration with Kerala Forest Research Institute (KFRI) and Kerala Forest Department. 18 p.

Anon. 1996. Asia-Pacific Water bird Conservation Strategy, 1996-2000. Wetlands International Asia-Pacific, Kula Lumpur, Publication No. 117, and International Waterflow and Wetlands Research Bureau-Japan Committee, Tokyo.

Anon. 2001. Vembanad Water Bird Count 2001 - a report. Kerala Forest Department and Kottayam Nature Society. 31 p.

Anon. 2003. Vembanad Water Bird Count 2003 - a report. Kerala Forest Department and Kottayam Nature Society. 80 p.

Balachandran, S. 1995. Shore birds of the Marine National Park in the Gulf of Mannar, Tamil Nadu. *J. Bombay nat. Hist. Soc.* 92(3): 303-313.

Balachandran, P.V., G. Mathew and K.V. Peter 2002. Wetland agriculture problems and prospects. pp. 50-68. In: Wetland Conservation and Management in Kerala (Ed.) Jayakumar. State Committee on Science Technology and Environment, Thiruvananthapuram.

Balagopalan, M. 1990. Soil and plant community relationships in wet evergreen forests of Silent Valley. In: Ecological Studies and Long-term Monitoring of Biological Processes in Silent Valley National Park. KFRI Research Report, 135-206.

Baker, H.R.1911. The occurrence of a Booby *Sula cyanops* at Cannanore. *J. Bombay Nat. Hist. Soc.* 21(1): 272-273.

Bell, H.L. 1983. A bird community of low land rainforest in New Guinea. 6. Foraging ecology and community structure of the avifauna. *Emu* 84(3): 142-159.

Bell, H.L. and S. Ferrier 1985. The reliability of estimates of density from transect counts. *Corella* 9: 3-13.

BirdLife International 2001. Threatened Birds of Asia. The BirdLife International Red Data Book. Cambridge, UK, BirdLife International.

Bourdillon, T.F. 1880. Letter to the Editor. *Stray Feather* 9(4) : 298-300.

Blacker, D. 1969. Behaviour of the Cattle Egret *Bubulcus ibis*. *Ostrich* 40: 74- 129.

Boecklen, W. J and D. Simberloff 1986. Area-based extinction models. In: Conservation in Dynamics of Extinction, (Ed.) Elliott, D.K., John Wiley and Sons, New York, 247-76.

Boomsa, J.J. and A.J. Van Loon 1982. Structure and diversity of ant communities in successive coastal dune valleys. *J. Anim. Ecol.* 51:957-974.

Bourdillon, T. F. 1880. Letters to the Editor. *Stray Feathers* 9(4): 298-300.

Bredin, D. 1983. Contribution a I etude ecologique d *Ardeola ibis* (L): heron garde boeufs de camargue. Ph.D. Dissertation, University Paul Sabatiesde Toulouse, Toulouse, France.

Brosset, A. 1990. A long term study of the rain forest birds in m'Passa (Gabon). In: A. Keast (Ed.) Biography and Ecology of Forest Birds SPB Academic, The Hague.

Burnham, K.P., D.R. Anderson and J.L. Laake 1980. Estimation of density from line transect sampling of biological populations. *Wildl. Monogr.* 72:202 P.

Burnham, K.P., D.R. Anderson and J.L. Laake 1981. Line transect estimation of bird population density using a Fourier series. In. Estimating the Number of Terrestrial Birds. (Eds.) C.J. Ralph and M.J.Scott. Studies in Avian Biology No. 6. Cooper Ornithological Society.

Caraco, T. 1980. On foraging time allocation in stochastic environment. *Ecology* 6: 119-128.

Cezilly, F. and J. Wallace 1988. The determination of prey captured by birds through direct field observations: a test of the method. *Colonial Waterbirds* 11(1): 110-112.

Champion, H.G. and S.K. Seth 1968. A Revised Survey of the Forest Types of India. Govt. of India. 404 p.

Chellappan, M. and J. Thomas 2001. Diversity of avifauna of the paddy fields of Pattambi, Palakkad District of Kerala. *National Symposium on Biodiversity Vis-à-vis Resource exploitation: An introspection, Port Blair*. 105 p.

Cody, M.L. 1970. Chilean bird distribution. *Ecology* 51: 445-464.

Cody, M.L. 1981. Habitat selection in birds: The roles of vegetation structure, competitors, and productivity. *Bioscience* 31(2): 107-113.

Cody, M.L. (Eds.) 1985. Habitat Selection in Birds. Academic Press, New York. 410 p.

Combs, D.L. and L.H. Fredrickson 1996. Foods used by male Mallards wintering in south eastern Missouri. *J. Wildl. Manage.* 60: 603-610.

Conner, M.A. 1979. Feeding association between Little Egret and African Spoonbill. *Ostrich* 50: 118.

Connors, P.G., J.P. Myers, C.S.W. Connors and F.A. Pitelka 1981. Inter habitat movements by Sanderling in relation to foraging profitability and tidal cycle. *Auk* 98(1): 49-64.

Cooper, J. 1985. Biology of Bank Cormorant part 3: foraging behaviour. *Ostrich* 56: 86-95.

Crawford, H.S., R.G. Hooper and R.W. Titterington 1981. Songbird population response to silvicultural practices in central Appalachian hardwoods. *J. Wildl. Manage.* 45: 680-692.

Custer, T.W. and R.G. Osborn 1978. Feeding site description of three heron species near Beaufort, North Carolina. pp. 355-360. In: Sprunt, A. Iv, J.C. Odgen and S. Winckler (Eds.), Wading Birds National. Audubon Society Research Report 7.

Daniels, R.J.R. 1989. A conservation strategy for the birds of the Uttara Kannada District. Ph.D. Thesis. Indian Institute of Science, Bangalore.

Drent, R.H. and S. Dann 1980. The prudent parents, energetic adjustments in avian breeding. *Ardea* 68: 225-252.

Driscoll, P. and M. Ueta 2000. Satellite tracking Eastern Curlews (*Numenius madagascariensis*) on northward migration from Moreton Bay and Western Port, techniques, flight, performance, migration avocets and strategies report for the Queensland Environment Protection Agency, on behalf of the Queensland Wader Study Group and the Queensland Ornithological Society.

Emlen, J.T. 1971. Population densities of birds derived from transect counts. *Auk* 88: 323-342.

Esler, D. 1992. Habitat use by Picivores birds on a Power Plant Cooling Reservoir. *J. Field Ornithol.* 63(3): 241-249.

Moss, D. 1978. Diversity of wood land song-bird populations. *J. Anim. Ecol.* 47: 521-527.

Evans, P.R. and P.J. Dugan 1984. Coastal birds: numbers in relation to food resources. pp. 8-29. *In*: Evans, P., J.D. Goss-Custard and W.G. Hale (Eds.) Coastal waders and wildfowl in winter. Cambridge University Press, London.

Faizi, S. 1985. An additional record of the Least Frigate Bird (*Fregata ariel*) in India. *J. Bombay nat. Hist. Soc.* 82(1): 191.

Fasola, M. 1982. Feeding dispersion in the Night Heron *Nycticorax nycticorax* and Little Egret *Egretta garzetta* and the information centre hypothesis. *Bull. Zool.* 49: 177-186.

Fasola, M. 1994. Opportunistic use of foraging resources by Heron communities in Southern Europe. *Ecography* 17: 113-123.

Fasola, M. and M. Ghidini 1983. Use of feeding habitat by breeding Night Heron and Little Egret. *Avocetta* 7(1): 29-36.

Fasola, M. and X. Ruiz 1996. The value of rice fields as substitutes for natural wetlands for waterbirds in the Mediterranean region. *Colonial Waterbirds* 19: 122-128.

Fasola, M., L. Canova and N. Saino 1996. Rice fields support a large portion of herons breeding in the Mediteranean region. *Colonial Waterbirds* 19: 129-134.

Ferguson, H.S., and T.F. Bourdillon, 1903. The birds of Pravamone with notes of their nidification, *J. Bombay Nat. Hist. Soc.* 15(2): 249-264.

Ferguson, H.S. and T.F. Bourdillon 1904a. The birds of Travancore. Part II. *J. Bombay nat. Hist. Soc.* 15(3): 654-673.

Ferguson, H.S. and T.F. Bourdillon 1904b. The birds of Travancore. Part II. *J. Bombay nat. Hist. Soc.* 16(1): 163-166.

Ferguson, H.S. and T.F. Bourdillon 1904c. The birds of Travancore. Part III. *J. Bombay nat. Hist. Soc.* 16(1): 1-18.

Ficken, M.A. and R.W. Ficken 1967. Age specific differences in the breeding behaviour of the American Redstart. *Wilson Bull.* 79: 88-199.

Fischer, R.A., A.S. Corbet and C.B. Williams 1943. The relationship between the number of species and the number of individuals in a random sample of an animal population. *J. Anim. Ecol.* 12: 42-58.

Francklin, J.F. 1989. Importance and justification of long-term studies in ecology. pp. 3-19. *In*: Long term Studies in Ecology (Ed.) G.E. Likkens. Springer-Verlag, New York.

Franzeb, K.E. 1981. The determination of avian densities using the variable-strip and fixed-width transect surveying methods. *Stud. Avian Biol.* 66: 6139-6145.

Fraser, M. 1974. Feeding association between Little Egret and Reed Cormorant. *Ostrich* 45: 262.

Fredrickson, L.H. and T.S. Taylor 1982. Management of seasonally flooded impoundments for wildlife. United States Fish and Wildlife Service Resource Publication 148 p.

Giller, P. S. 1984. Community structure and the Niche, Chapman and Hall, 58 p.

Gokula, V. 1998. Bird communities of the thorn and dry deciduous forests in Mudumalai Wildlife Sanctuary, South India. Ph.D. Thesis. Bharathiar University, Coimbatore. 200 p.

Gokula, V. and L. Vijayan 1996. Birds of Mudumalai Wildlife Sanctuary. *Forktail* 12: 107-117.

Gomez, K.A. and A.A. Gomez 1984. Statistical Procedures for Agricultural Research. John Wiley & Sons, New York. 680 p.

Gopalan, U.K. 1991. Backwaters: Our Valuable Resources (in Malayalam), Kerala Sasthra Sahitya Parishad, Kozhikode. 48 p.

Goss-Custard, J.D. 1977. The ecology of the Wash III. Density related behaviour and the possible effects of loss of feeding grounds on wading birds (Charadrii). *J. Appl. Ecol.* 14: 721-739.

Goss-Custard, J.D. 1979. The energetics of foraging Red Shank *Tringa totanus. Stud. Avian Biol.* 2: 247-257.

Goss-Custard, J.D. and M.G. Yates 1992. Towards predicting the effect of salt marsh reclamation on feeding birds numbers on the wash. *J. Appl. Ecol.* 29: 330-340.

Grant, P.R. and B.R. Grant 1987. The extraordinary El Nino event of 1982-1983: effects on Darwin's finches on *Isla Genovesa*, Galapagos. *Oikos* 49: 55-56.

Greenberg, R. and J. Gradwohl 1986. Constant density and stable territoriality in some tropical insectivorous birds. *Oecologia* 69: 618-625.

Grimme H., R., C. Inskipp, T. Isskipp. 1988. Birds of Indian Subcontinent. Oxford University Press. 888 p.

Gudmundsson, A.G. and A Gardarsson 1993. Numbers, geographic distribution and habitat utilization of waders (Charadrii) in spring on the shores of Iceland. *Ecography* 16(1): 82-93.

Hafner, H. and R.H. Britton 1983. Changes of foraging sites by nesting Little Egret (*Egretta garzetta* L.) in relation to food supply. *Colonial Waterbirds* 6: 24-30.

Hafner, H., V. Boy and G. Gorl, 1982. Feeding methods, flock size and feeding success in the Little Egret *Egretta garzetta* and Squacco Heron *Ardeola ralloides* in Camorgue, Southern France. *Ardea* 70: 45-54.

Hafner, H., P.J. Dugan and Vincent Boy 1986. Use of artificial and natural wetlands as feeding sites by Little Egret (*Egretta garzetta L.*) in the Camarge, Southern France. *Colonial Waterbirds* 9(2): 149-154.

Herrera, C. 1988. Variaciones anuales en las poblaciones de pajaros frugivores Y su relation con la abundancia de frutos. *Ardeola* 35: 135-142.

Hilden, O. 1981. Sources of error involved in the Finnish line transect method. Stud. *Avian Biol.* 6: 152-159.

Hoffmann, T. W. 1983. Occurrence of certain waders in Sri Lanka. *J. Bombay nat. Hist. Soc.* 79: 668-669.

Holmes, R.T. and F.A. Pitelka 1968. Food overlap among coexisting Sandpiper on Northern Alaskan Tundra. *Syst. Zool.* 17: 305-318.

Holmes, V.R., T.T. Cable and V. Brack Jk 1986. Avifauna as indicators of North Indiana. *Indian Academy of Science (Zool.)* 95: 523-528.

Hoves, J.G. and D. Bakewell 1989. Shore bird studies manual AWB. Publications No. 55 Kula Lumpur. 362 p.

Howe, R. W., T.D. Howe and H.A. Ford 1981. Bird distributions on small rain forests remnants in New South Wales. *Australian Wildl. Res.* 8(3): 637-652.

Hume, A.O. 1876. A first list of the birds of the Travancore hills. *Stray Feathers* 4(4, 5 & 6): 351-405.

Hume, A.O. 1878. A second list of the birds of southern Travancore. *Stray Feathers* 7(1-2): 33-39.

Hurlbert, S.H. 1971. The non-concept of species diversity: A critic and alternative parameters. *Ecology* 52: 577-86.

Hurlbert, S.H. 1978. The measurement of niche overlap and some relatives. *Ecology* 59: 67-77.

Hyman, L. 1967. The Invertebrates, Vol. VI. Molusca I. McGraw Hill Book Co., New York. 792 p.

Jackson, M.C.A. 1954a. Occurrence of the Banded Crake *Rallus eurizonoides amuroptera* (Jerdon) in Trvancore. *J. Bombay nat. Hist. Soc.* 52(1): 211-212.

Jaekson, M.C.A. 1954b. The occurrence of Franklins Nightjar *Caprimvigvs monticows monticolvs* in Travancore-Cochin. *J. Bombay nat. Hist. Soc.* 52(2&3): 603.

Jairaj, A.P. and V.K.S. Kumar 1990. Occurrence of Spoonbill *Platalea leucorodia* Linn. in Kerala. *J. Bombay nat. Hist. Soc.* 87(2): 289.

James, E.J. 1983. Establishing a hydro-metrological data bank at CWRDS. Proc. National Workshop on Scientific methods to collection and documentation of hydrologic data. CWRDM, Kozhikode, Kerala. pp. 29-35.

James, F.C. and N.O. Wamer 1982. Relationships between temperate forest bird communities and vegetation structure. *Ecology* 63: 159-171.

Jayson, E.A. 1989. Seasonal abundance of avifauna in Silent Valley National Park, Kerala. Abstract of the National Symposium on Recent Advances on Behavioural Sciences. Dept. of Zoology, University of Rajasthan, Jaipur.

Jayson, E.A. 1994. Synecological and behavioural studies on certain species of forest birds. Ph.D. Thesis. University of Calicut, India. 314 p.

Jayson, E.A. 2000. Occurrence of Black Stork *Ciconia nigra* (Linnaeus) in Kole Wetlands of Thrissur, Kerala. *Newsl. Birdwatchers* 40(3): 39-40

Jayson, E.A. and D.N. Mathew 1993. Diversity and species-abundance distribution of birds in the adjoining forests of Silent Valley National Park. Proceedings of the Fifth Kerala Science Congress. 80-83.

Jayson, E.A. and P. V. Nair 1993. Wildlife and land use pattern of Malayattoor. In. Baseline studies for the proposed nature study centre at Kalady in the Malayattoor Forest Division (Eds). Basha C. and K.K.N., Nair. *Kerala Forest Research Institute,* Peechi. pp. 111-134.

Jayson, E.A. and C. Sivaperuman 1999. Kole lands of Thrissur: A threatened wetland ecosystem. *Evergreen* 43: 10-11.

Jayson, E A. and P. S. Easa 2000. Documentation of vertebrate fauna in Mangalavanam mangrove area. KFRI Research Report No. 183. *Kerala Forest Research Institute,* Peechi. 42 p.

Jayson, E.A. and D.N. Mathew 2000a. Seasonal changes of tropical forest birds in southern Western Ghats. *J. Bombay nat. Hist. Soc.* 97: 52-61.

Jayson, E.A. and D.N. Mathew 2000b. Diversity and species abundance distribution of birds in the tropical forests of Silent Valley, Kerala. *J. Bombay nat. Hist. Soc.* 97: 52-61.

Jayson, E.A. and D.N. Mathew 2002. Structure and composition of two bird communities in the Southern Western Ghats. *J. Bombay nat. Hist. Soc.* 99 (1): 8-25.

Jayson, E.A. and C. Sivaperuman 2002. Seasonal variation of wetland birds in the Kole lands of Thrissur Kerala. Proceedings of the National Seminar on Ecology and Conservation of wetlands, 31st January - 2nd February 2002, Christ College, Irinjalakuda, Thrissur. pp. 29-38.

Jayson, E.A. and C. Sivaperuman 2003. Sighting of Lesser Frigate Bird *Fregata ariel* Gray in the Kole Wetlands of Thrissur, Kerala. *J. Bombay nat. Hist. Soc.* 100(1): 106-108.

Jayson, E.A. and D.N. Mathew 2003. Vertical stratification and its relation to foliage in tropical forest birds in Western Ghats (India). *Acta Ornithol.* 38: 111-116.

Jayson, E.A. and C. Sivaperuman 2006. Diversity of birds in the tropical forests of Kerala, India. National Conference on Forest Biodiversity Resources: Exploitation, Conservation and Management. 21st - 22nd March 2006. pp. 114.

Jhunjhunwala, S., A.R. Rahmani, F. Ishtiaq and Z. Islam 2001. The Important Bird Area programme in India. *Buceros* 6(2): 1-49.

Jin-Han Im, Byund-Ho Yoo, Changman Won, Jin-Young Park and Jeong-Yeon Yi 2001. An agricultural habitat indicator for wildlife, OECD Expert meeting on Agri-Biodiversity Indicator, November 2001, Switzerland. 12 p.

Johnkutty, I. and V.K. Venugopal 1993. Kole Lands of Kerala. Kerala Agricultural University. 68 p.

Johnsingh, A.J.T, and J. Joshua 1994. Avifauna in three vegetation types on Mundathurai plateau, south India. *Journal Tropical Ecology* 10: 323-335.

Johnsingh, A.J.T, M.H. Martin, J. Balasingh and V. Chelladurai, 1987. Vegetation and avifauna in a thorn scrub habitat in South India. *Trop. Ecol.* 28: 22-34.

Kalejta, B. 1991. Aspects of ecology of migrant shorebirds (Aves: Charadrii) at the Berg River Estuary, South Africa. Ph.D. Thesis. University of Cape Town, South Africa.

Kalejta, B. 1993. Diets of shorebirds at the Berg River Estuary, South Africa: spatial and temporal variation. *Ostrich* 64: 123-133.

Kalejta, B. and P.A.R. Hockey 1994. Distribution of shorebirds at the Berg River Estuary, South Africa, in relation to foraging mode, food supply and environmental features. *Ibis* 136: 233-239.

Karr, J.R. 1971. Structure of avian communities in selected Panama and Illionis habitats. *Ecol. Monogr* 41: 207-233.

Karr, J.R. 1980. Geographical variation in the avifauna of tropical forest undergrowth. *Auk* 97: 283-298.

Karr, J.R. and R.R. Roth 1971. Vegetation structure and avian diversity in several New World areas. *Amer. Nat.* 105: 423-435.

Katti, M.V. 1989. Bird communities of Lower Dachigam Valley, Kashmir. M.Sc. Dissertation. Saurashtra University, Ragkot.

Katzir, G., T. Stord, E. Schechtman, S. Harelis and Z. Arad 1999. Cattle Egrets are less able to cope with light refraction than are other heron. *Anim. Behaviour* 57: 687-694.

Kent, D.M. 1986. Foraging behaviour habitat use and food of three egrets in a marine habitat. *Colonial Waterbirds* 9.

Koen, K.H. 1992. Medium-tern fluctuations of birds and their potential food resources in the Knysna forest. *Ostrich* 63: 21-30.

Krebs, C.J. 1991. The experimental paradigm and long-term studies. *Ibis.* 133(Suppl.) 1: 3-8.

Krebs, J.R. and A. Kacelnik 1991. Decision making. In: Behavioural Ecology: an evolutional approach (Eds.): J.R. Krebs and N.B. Davies. pp. 105-136. Blackwell Scientific Publications, Oxford.

Kumar, P.M. and P.M. Kumar 1996. Occurrence of Christmas Frigate Bird. *Newsl. Birdwatchers* 36(6): 113-114 .

Kumar, S. 1990. Blacktailed Godwit (*Limosa limosa*) and Large Indian Pratincole (*Glareola pratincola*) - two new records from Kerala. *J. Bombay nat. Hist. Soc.* 87(2): 296.

Kurup, D.N. 1989. Sight records of Ibis and storks in Kerala. *J. Bombay nat. Hist. Soc.* 86(2): 239.

Kurup, D.N. 1991. Ecology of the birds of Malabar Coast and Lakshadweep. Ph.D. Dissertation, University of Calicut, Calicut.

Kurup, D.N. 1996. Ecology of the birds of Barathapuzha estuary and survey of the coastal wetlands of Kerala. Final Report submitted to Kerala Forest Department, Trivandrum. 59 p.

Kushlan, J.A. 1976. Feeding Behaviour of North American heron. *AUK* 93: 86-94.

Kushlan, J.A. 1978. Feeding ecology of wading birds. pp. 242-297. In: 'Wading Birds' Sprunt, A., *et al.* (Eds.) National Audubon Society, New York.

Kushlan, J.A. 1986. Response of wading birds to seasonality fluctuating water levels: strategies and their limits. *Colonial Waterbirds* 9: 155-162.

Kushlan, J.A. 1987. Extreme threats and internal management : The hydrologic regulation of the Everglades, Florida, USA. *Environ. Manage* 11: 109-119.

Lamba B.S. 1970. Black winged kite Elams carleus vocifens (Latham) taking in flight wounded pigeon. Treon phoenicoptera (Lathan). *J. Bombay nat. Hist. Soc.* 66(3): 22.

Lambshead, P.J.D., G.L.J. Paterson and J.D. Gage 1997. BioDiversity Professional Beta. The Natural History Museum and The Scottish Association for Marine Science.

Landers, P.B. and J.A. MacMahon 1980. Guilds and community organization: Analysis of an oak woodland avifauna in Sonora, Mexico. *Auk* 97: 351-365.

Lanes, J. and M. Fujioka 1998. The impact of changes in irrigation practices on the distribution of foraging egrets and herons (Ardeidae) in the rice fields of Central Japan. *Biol. Conserv.* 83: 221-230.

Levins, R. 1968. Evolution in changing environments: some theoretical exploration, Princeton University Press, Princeton, JN.

Lovejoy, T.E. 1975. Bird diversity and abundance in Amazon forest communities. *Living Bird* 13:127-191.

Ludwig, J. A. and J. F. Reynolds 1988. Statistical Ecology, A premier on Methods and Computing. A Wiley-Interscince publication. 337 p.

Mac Arthur, R.H. and J.W. Mac Arthur, 1961. On bird species diversity. *Ecology* 50: 793-801.

Mac Arthur, R.W., J.W. Mac Arthur, and J. Preer. 1962. On bird species diversity. II prediction of bird census from habitat measurements. *Am. Nat.* 96: 167-174.

Mac Arthur, R.A. 1972. Geographical Ecology. Harper and Row, New York.

Mac Arthur. R.H, H. Recher and M.L. Cody 1966. On the relation between habitat selection and bird species diversity. *Am. Nat.* 100: 319-332.

MacArthur, R.H. 1958. Population ecology of some Warblers of North eastern Coniferous forest. *Ecology* 39: 599-619.

MacArthur, R.H. and E.R. Pianka 1966. On the optimal use of patchy environment. *Am. Nat.*100: 603-609.

Magurran, A.E. 1988. Ecological Diversity and its Measurement. Croom Helm Ltd., London. 179 p.

Manilal, K.S. 1988. Flora of Silent Valley Tropical Rainforests of India. The Mathrubhumi (MM) Press, Calicut. 398 p.

Miranda, L. 1995. Wading birds predation in tropical mangrove swamps implications to Juvenile fish population dynamics, M.S. Thesis, North Carolina State University, Raligh, North Carolina.

Montogomery, D.C. 1991. Design and Analysis of Experiments. John Wiley and Sons, New York. 649 p.

Morales, G. and J. Pacheco 1986. Effects of diking of Venezuelan Savana on avian habitat, species diversity, energy flow, and minerals flow through wading birds. *Colonial Waterbirds* 9: 236-242.

Morris, F.T. 1978. Feeding association between Little Egret and Sacred Ibis. *Emu* 78: 164.

Morrison, M. L., A. Kimberly and I. C. Timossi 1980. The structure of a forest bird community during winter and Summer. *Wilson Bull.* 98(2): 214-230.

Morrison, M.L., B.G. Marcot and R.W. Mannan (Ed.) 1998. Wildlife-habitat Relationships. University of Wisconsin Press, Madison.

Moss, D. 1978. Diversity of wood land song. bird poforlation. *J. Anim. Ecol.* 47: 521-527.

Moser, M. E. and R. W. Summer 1987. Wader population of the non-estuarine coasts of Britain and Northern Ireland; Results of the 1984-85 Winter Shorebird Count. *Bird Study* 34: 71-81.

Mukherjee, A.K. 1969. Food-habits of waterbirds of the Sundarban, 24 Parganas District, West Bengal, India. *J. Bombay nat. Hist. Soc.* 66(2): 345-360.

Mukherjee, A.K. 1972. Food-habits of waterbirds of the Sundarban, 24 Parganas District, West Bengal, India-III. Egrets. *J. Bombay nat. Hist. Soc.* 68 (3): 691-716.

Muller, H.C., M.G.B. Biban and H.F. Scars 1972. Feeding interactions between Pied-billed Cranes and Herons. *Auk* 189-190.

Myers, J.P. 1984. Spacing behaviour of non-breeding shorebirds. *Behav. Marine Anim.* 6: 272-323.

Myers, J.P., S.L. Williams and F.A. Pitelka 1980. An experimental analysis of prey availability for Sanderling (Aves: Scolopacidae) feeding on sandy beach crustaceans. *Can. J. Zool.* 58: 1564-1574.

Myers, J.P., R.I.G. Morrison, P.Z. Anta, B.A. Harrington, T.E. Lovejoy, M. Sallaberry, S.E. Senner and A. Tarak 1987. Conservation strategy for migratory species. *Am. Sci.* 75: 19-26.

Nagarajan, R. and K. Thiyagesan 1996. Waterbirds and substrate quality of the Pichavaram wetlands, Southern India. *Ibis* 138: 710-721.

Nagarajan, R. and K. Thiyagesan 1998. Significance of adjacent croplands in attracting waterbirds to the Pichavaram mangrove forests. pp. 172-181. In: Birds in Agricultural Ecosystem. (Eds) Dhindsa, M.S., P.S. Rao and B.M. Parasharya. Society for Applied Ornithology, Hyderabad.

Nair, S.C. 1988. Long-term Conservation Potential of Natural Forests in the Southern Western Ghats of Kerala. Report submitted to the MAB committee, 323 p.

Namasivayam, L. and R. Venugopalan 1990. Avocet *Recurvirostra avocetta* in Kerala. *J. Bombay nat. Hist. Soc.* 86(3): 447.

Neelakandan K.K. 1958. "Keralathilae Pakshikal" Kerala Sahitya Academy, Trichur. 529 p.

Neelakantan, K.K. 1968: A note on the behaviour of two House Crows. *Newsletter for Birdwatchers* 8(9): 2-3.

Neelakantan, K.K. 1969. Motiveless malignity or purposeless pestering. *Newsl. Birdwatchers* 9(6): 4.

Neelakantan, K.K. 1970. The occurrence of the Sanderling *Calidris alba* Kerala. *J. Bombay nat. Hist. Soc.* 67(3): 570.

Neelakantan, K.K. 1990. Breeding of the River Tern *Sterna aurantia* in Kerala. *J. Bombay nat. Hist. Soc.* 87(1): 144-145.

Neelakantan, K.K. and V.K. Sureshkumar 1981. Occurrence of the Blackwinged Stilt *Himantopus himantopus* in Kerala. *J. Bombay nat. Hist. Soc.* 77(3): 510.

Neelakantan, K.K., K.V. Sreenivasan and V.K. Sureshkumar 1981. The Crab Plover *Dromas ardeola* in Kerala. *J. Bombay nat. Hist. Soc.* 77(3): 508.

Neelakantan, K.K., C. Sasikumar and R. Venugopalan 1993. A Book of Kerala Birds. World Wide Fund for Nature-India, Kerala State Committee, Trivandrum 146 p.

Newton, I. 1998. Population Limitation in Birds. Academic Press, New York.

Nilson, S.G. 1979a. The effect of forest management on the breeding bird community in southern Sweden. *Biol. Conserv.* 16: 135-143.

Nilson, S. G. 1983. The structure of bird communities of natural forests in Sweden and the grading success of birds nesting in natural cavities. *Ibis* 125(4): 587-588.

Nilsson, L.G. and I.N. Nilsson 1978. Breeding bird community densities and species richness in Lakes. *Oikos* 31: 219-221.

Odum, E. P. 1974. Fundamentals of Ecology. W.B. Saunders company London. 144 p.

Owen, M. and J.M. Black 1991. Geese and their future fortune. *Ibis Suppl.* 133(2): 28-35.

Pain, D.J. and M.W. Pienkowski (Eds.) 1996. Farming and birds in Europe: the common agricultural policy and its implications for bird conservation. Academic Press, London.

Pandey, D.J. 1958. Cormorants and Egrets fishing in cooperation. *J. Bombay nat. Hist. Soc.* 55: 170-171.

Parker, G.R., M.J. Petrie and D.T. Sears 1992. Waterfowl distribution relative to wetland acidity. *J. Wildl. Manage.* 56: 268-274.

Parks, J.M. and S.L. Bressler 1963. Observations of joint feeding activities of certain fish eating birds. *Auk* 80: 198-199.

Pearson, D. 1977. A pantropical comparison of bird community: Structure on six low land rainforests sites. *Condor* 79: 232-244.

Pearson, D.L. 1982. Historical factors and bird species richness. In. Biological Diversification in the Tropics. Ed. G.T. Prance. New York, Columbia University Press. 441-452.

Peet, R.K. 1974. The measurement of species diversity. *Ann. Rev. Ecol. Systems* 5: 285-307.

Petraitis, P.S. 1979, Likelihood measures of niche breadth and overlap. *Ecology* 60: 703-710.

Pielou, E.C. 1975. Ecological Diversity. A Wiley-Interscience publication, John Wiley and Sons, New York.

Praveen, J. and M. Kumar 1996. Glossy Ibis in Palakkad District. *Newsl. Birdwatchers* 36(1): 13.

Preston, F.W. 1948. The commonness and rarity of species. *Ecology* 29:254-283.

Preston, F.W. 1962. The canonical distribution of commonness and rarity. *Ecology* 43: 185-215; 410-432.

Price, T.D. 1979. The seasonality and occurrence of birds in the Eastern Ghats of Andhra Pradesh. *J. Bombay nat. Hist. Soc.* 76(3): 379-422.

Prime Rose, J.B. 1904. Birds observed in the Nilgiris and Wayanad. *J. Bombay nat. Hist. Soc.* 16(1): 163-166.

Pulliam, H.R. 1980. Do chipping Sparrows forage optimally? *Ardea* 68: 75-82.

Puttick, G.M. 1981. Sex related difference in the foraging behaviour of Curlew Sandpiper. *Ornis. Scand.* 12: 13.

Pyke, G.H. 1984. Seasonal patterns of abundance of insectivorous birds and flying insects. *Emu* 85(1): 34-39.

Ramakrishnan, P. 1983. Environmental Studies on the Birds of Malabar Forest. Ph. D. Dissertation, University of Calicut, Calicut.

Ramo, C. and B. Busato 1993. Resource use by Herons in Yucatan wetland during the breeding season. *Wilson Bull.* 105: 573-586.

Rao, G.N. 1983. Statistics for Agricultural Sciences. Oxford and IBH Publishing Company.

Ravindran, P.K. 1993. Occurrence of the Glossy Ibis in Kole wetland, Thrissur District Kerala. *Newsl. Birdwatchers* 33(6): 109.

Ravindran, P.K. 1994. Gadwall at Kadalundy Estuary, Kerala. *Newsl. Birdwatchers* 34(2): 33.

Ravindran, P.K. 1995. The Kole Wetlands - an avian paradise in Kerala. *Newsl. Birdwatchers* 35(1): 2-5.

Ravindran, P.K. 1998. Sighting of the Comb Duck in Kerala. *Newsl. Birdwatchers* 38(4): 71.

Ravindran, P.K. 1999. Whitenecked Stork in Kole wetlands. *Newsl. Birdwatchers* 39(3): 51.

Ravindran, P.K. 2001. Occurrence of the White-winged Black Tern *Chlidonias leucopterus* in Kerala. *J. Bombay nat. Hist. Soc.* 98(1): 112-113.

Ray, J.G. 1992. The Watercock *Gallicrex cinerea* in Kuttanadu, Kerala. *J. Bombay nat. Hist. Soc.* 88(2): 283.

Recher, H.F. 1969. Bird species diversity and habitat diversity in Australia and North America. *Am. Nat.* 103: 75-80.

Recher, H.F. and J. A. Recher 1972. The foraging behaviour of the Reef Heron. *Emu* 72: 85-90.

Reynolds, J. 1965. Association between Egrets and African Spoonbill. *Brit. Birds* 58: 468.

Robin, V.V. and P. Davidar 2002. The vertical stratification of birds in mixed species flocks at Parambikulam, South India: A comparison between two habitats. *J. Bombay Nat. Hist. Soc.* 99(3): 389-399.

Rodger, W.A. and H.S. Panwar 1989. Planning a Wildlife Protected Area Network in India. Wildlife Institute of India, Dehra Dun. Vol. 2. 264 p.

Rosenberg, R. 1976. Benthic faunal dynamics during succession following pollution abatement, in a Swedish estuary. *Oikos* 27: 414-27.

Saikia, P. and P.C. Bhattacharjee 1990b. Conservation of large wetland birds in Assam: Adjutant Stork. pp. 50-51. In: Seminar on Wetland Ecology and Management. Bombay Natural History Society, Bombay.

Sampath, K. 1989. Studies on the ecology of Shorebirds (Aves: Charadriiformes) of the Great Vedaranyam Swamp and the Pichavaram Mangroves of India. Ph.D. Dissertation, Annamalai University, Tamil Nadu. 202 p.

Sampath, K and K. Krishnamurthy 1989. Shore birds of the Salt ponds at Great Vedaranyam Salt Swamp. *Stilt* 15: 20-23.

Sampath, K. and K. Krishnamurthy 1990. Bird fauna and limnology of the Koliveli tank, Tamil Nadu. pp. 47-48. In: Vijayan, V.S. (Ed.) Proc. Seminar on wetland Ecology and management. Bombay Natural History Society, Bombay.

Sampath, K., K. Krishnamurthy and V.S. Vijayan 1995. Foraging behaviour of Shorebirds. *J. Ecol. Soc.* 8: 13-22.

Santharam, V. 1995. Ecology of sympatric Woodpecker species of Western Ghats, India. Ph.D. Dissertation, Pondicherry University, Pondicherry, India. 162 p.

Saraswathy, U. 1999. Lesser Frigate Bird *Fregata minor aldabrensis* Mathews on the Kerala coast. *J. Bombay nat. Hist. Soc.* 96(2): 313.

Sashikumar, C. 1992. Occurrence of the Indian Shag *Phalacrocorax fuscicollis* Stephens in Kerala. *J. Bombay nat. Hist. Soc.* 88(3): 442.

Sathasivam, K. 1992. Painted Stork *Mycteria leucocephala* (Pennant) in Kerala. *J. Bombay nat. Hist. Soc.* 89(2): 246.

Satheesan, S.M. 1990. The ecology and behaviour of the Pariah Kite (*Milvus migrans govinda*) Sykes as a problem bird at some Indian aerodromes. Ph.D. Dissertation, University of Bombay, Bombay, India. 248 p.

Schemske, D.W. and N. Brokaw 1981. Tree falls and the distribution of under story birds in a tropical forest. *Ecology* 62(4): 938-945.

Schoener, T.W. 1971. Theory of feeding strategies. *Ann. Rev. Ecol. Syst.* 2: 369-403.

Shufford, W.D., G.W. Page and J.E. Kjelmyr 1998. Patterns and dynamics of shorebird use of California's Central Valley. *Condor* 100: 227-244.

Shugart, H.H., JR., and D.L. Urban 1986. Modelling habitat relationship of terrestrial vertebrates: the researchers view point. pp. 425-429. In: Verner, J., M.L. Morrison and C.J. Raiph (Eds.) Wildlife 2000: Modelling habitat relationship of terrestrial vertebrates. University of Wisconsin Press, Wisconsin USA.

Sivaperuman, C. and E.A. Jayson 2001a. Structure and species composition of wetland birds in the Kole lands of Thrissur, Kerala. Proceedings of the Thirteenth Kerala Science Congress, Jan. 2001, Thrissur. pp. 152-155.

Sivaperuman, C. and E.A. Jayson 2001b. Diversity of wetland birds in the Kole lands of Thrissur Kerala. National seminar on Biodiversity conservation: Challenges for 21st Century, Nov. 2001, Gwalior. 44 p.

Sivaperuman, C. and E.A. Jayson 2002a. Occurrence of Northern Shoveller *Anas clypeata* Linnaeus in Kole Wetlands of Thrissur, Kerala. *J. Bombay nat. Hist. Soc.* 99(3): 517.

Sivaperuman, C. and E.A. Jayson 2002b. Population dynamics of waders (Charadriiformes) in the Kole lands of Thrissur, Kerala. Proceedings of Fourteenth Kerala Science Congress, Jan. 2002, Ernakulam. pp. 781.

Sivaperuman, C. and E.A. Jayson 2003. Habitat utilization of wetland birds the Kole wetlands of, Kerala, India. 28th conference of the Ethological Society of India, February 2003, Mundanthurai, Tirunelveli, Tamil Nadu. pp. 87-92.

Sjoberg, K. 1989. Time related predator/prey interactions between birds and fish in northern Swedish river. *Oecologia* 67: 53-39.

Skagen, S.K. and F.L Knopf 1994. Migrating shorebirds and habitat dynamics at a Prairie wetland complex. *Wilson Bull.* 106(1): 91-105.

Slocomb, J., B. Stauffer and K.L. Dickson 1977. On fitting the truncated lognormal distribution to species abundance data using maximum likelihood estimation. *Ecology* 58: 693-696.

Sodhi, N.S. 1986. Feeding ecology of Indian Pond Heron and its comparison with that of Little Egret. *Pavo* 24(1&2): 97-112.

Sodhi, N.S. and S. Khera 1984. Food requirement during growth and feeding behaviour of nestling *Bubulcus ibis coromandus* (Boddaert). *Pavo* 22(1&2): 21-39.

Soule, M.E. and B.A. Wilcox (Eds.) 1980. Conservation biology: An evolutionary ecological perspective. Sinauer Associate, Sunderland, Masachusetts, USA.

Steele, B.B. 1992. Habitat selection by breeding Black-throated Blue Warbler at two spatial scales. *Ornis Scandinavica* 23: 33-42.

Stephens, D.W. and J.R. Krebs 1987. Foraging Theory. Princeton University Press, Princeton, New Jersey.

Stiles, F.G. 1978. Temporal organization of flowering among the Humming Bird, food plants of a tropical forest. *Biotropica* 10: 194-210.

Strin, J. 1981. Manual of methods in Aquatic Environment Research. Part 8 Ecological Assessment of pollution effects (Guidelines for the FAO/GFCM)/ UNEP Joint coordinated project on pollution in the Mediterranean) FAO. Fish Tech. Pap. 209: 70 p.

Summer, R.W. and B. Kalejta-Summer 1996. Seasonal use of sand flats and salt marshes by waders at low and high tide at Langebann Lagoon, South Africa. *Ostrich.* 67: 62-72.

Sundaramoorthy, T. 1991. Ecology of terrestrial birds in Keoladeo National Park, Bharatpur. Ph.D. Thesis. University of Bombay.

Terborgh, J. 1985. Habitat Selection in Amazonian birds In: Habitat selection in birds. (Ed.) M.L. Cody, Academic Press, New York, 311-338.

Terborgh, J., S. K. Robinson, T.A. Parker III Charles A. Munn and N. Pierpont 1990. Structure and organization of an Amazonian forest bird community. *Ecol. Monogr.* 213-238.

Thiollay, J.M. 1986. Structure comparee du peuplement avien dans trois sites de foret primaire en Guyan. *Revue d'Ecologie la Terre et la Vie* 41:59-105.

Udvardy, M.D.R. 1975. A classification of the biogeographical provinces of the world. IUCN Occasional Paper. 18 IUCN, Gland, Morges.

Usher, M.B. 1986. Wildlife conservation evaluation: Attributes, criteria and values. In: Usher M.B. (Ed.). Wildlife Conservation Evaluation. (1986) Chapman and Hall, London. 3-44.

Uthaman, P.K. 1990. Spotbill Duck *Anas poecilorhyncha* J.R. Forester in Kerala. *J. Bombay nat. Hist. Soc.* 87(2): 290-291.

Uthaman, P.K. and L. Namasivayam 1991. The birdlife of Kadalundi Estuary. *Blackbuck* 7(1): 3-11.

Verma, A., N. Chaturvedi, S. Balachandran and I. Kehimkar 2002. Avian diversity in and around Mangroves and Mahul Creeks Mumbai, India. Proc. The National Seminar on Creeks, Estuaries and Mangroves - pollution and conservation. pp. 266-275.

Verner, J., and T.A. Larson 1989. Richness of breeding bird species in mixed conifer forests of the Sierra Nevada, California. *Auk* 106: 447-463.

Vijayagopal, K. 1991. Comparative biology and ecology of the Redwattled Lapwing *Vanellus indicus* (Boddaert) and the Yellowwattlled Lapwing *Vanellus malabaricus* (Boddaert) and a preliminary comparative study of the vocalisations of certain species of Indian birds. Ph. D. Thesis, University of Calicut, Kerala, India. 296 p.

Vijayan, L. 1984. Comparative biology of Drongos (Family: Dicruridae class: Aves) with special reference to ecological isolation. Ph. D. Thesis, University of Bombay.

Vijayan, V.S. and M. Balakrishnan 1977. Impact of Hydro-Electric Project on Wildlife, Report of the first phase of study, Kerala Forest Research Institute, Peechi. 111 p.

Ward, H.B. and G.C. Whipple 1945. Freshwater Biology. Chapman and Hall Limited, London.

Weller, M.M. 1978. Management of freshwater marshes for wildlife. In: Good, R.E., D.F. Whingham and R.L. Simpson (Eds.) Fresh Water Wetlands: Ecological process and management potential, New York. Academic Press.

Weller, M.W. 1994. Seasonal dynamics of birds assemblages in a Texas Estuarine Wetland. *J. Field Ornithol.* 65(3): 388-401.

White, G.C. and R.A. Garrott 1985. Habitat analysis: Analysis of Wildlife Radio tracking data. pp. 183-205.

Whitecomb, R.F., C.S. Robins, J.F. Lynch, B.L. Whitcomb, M.K. Klikiewicz and D. Bystrak 1981. Effects of forest fragmentation on avifauna of the eastern deciduous forest. pp. 125-206. In: R.L. Burgess and D.M. Sharpe (Eds.) Forest Island dynamics in man dominated landscapes. Springer Verlag. New York. USA.

Wiens, J.A. 1983. Avian Community ecology: An iconoclastic view In: Perspectives in Ornithology. A.H. Brush and G.A. Clark, (Eds.) Cambridge University Press, Cambridge, 335-403.

Wiens, J.A. 1984. Resource system, populations and communities. P.W. Pricen, C.N. Slobdchikoff and W.S. Gaud (Eds.). A new ecology: Novel approaches to interactive systems, John Wiley and Sons Inc.

Wiens, J.A. and J.T. Rottenberry 1981. Habitat associations and Community structure of birds in shrub steppe environments. *Ecol. Monogr.* 51: 21-41.

Wiens, J.A. 1989. The Ecology of Bird Communities Vol. I. Foundations and Patterns. Cambridge University Press, Cambridge. pp. 539.

Willard, D.E. 1977. The feeding ecology and behaviour of five species of herons, Southern New Jersey. *Condor* 79: 462-470.

Wilson, E.O. 1985. The biological diversity crisis. *Bioscience* 35: 700 706.

Winkler, H. 1983. The ecology of Cormorants. pp. 193-199. In: Limnology of Parakrama Samudra Sri Lanka. (Ed): F. Schiemer Junk, The Hague.

Wright, J.S. 1970. Competition between insectivorous lizards and birds in Central Panama. *Amer. Zool.* 19: 1145-1156.

Yahya, H.S.A. 1980. A comparative study of ecology and biology of barbets, *Megalaima* spp. (Capitonidae: Piciformes) with special reference to (*Megalaima viridis*) (Boddaert) and (*M. rubricapilla malabarica*) (Blyth) at Periyar Tiger Reserve, Kerala. Ph.D. Dissertation, University of Bombay, Bombay. 210 p.

Zacharias, V.J. 1979. Ecology and biology of certain species of Indian Babblers (*Turdoides* spp.) in Malabar. Ph.D. Dissertation, University of Calicut, Kerala, India. 196 p.

Zacharias, V.J. and A.J. Gaston 1999. The recent distribution of endemic, disjunct and globally uncommon birds in the forest of Kerala, State, South-west India. *Bird Conservation International* 9: 191-225.

Zwarts, L. 1978. Intra and inter specific competition for space in estuarine bird species in a one prey situation. *International Ornithological Congress* 17: 1045-1050.

Zwarts, L. 1985. The winter exploitation of Fiddler Crabs *Uca tangeri* by Waders in Guinea-Bissau. *Ardea* 73: 3-12.

❐❐❐

Index

C

D

N

O

P

Q

R

S

T

□□□